TELECOMMUNICATIONS

Equipment Fundamentals and
Network Structures

TELECOMMUNICATIONS

Equipment Fundamentals and Network Structures

VINCE COUGHLIN

Manager Network Support
Citicorp International Communications, Inc.

VNR VAN NOSTRAND REINHOLD COMPANY

Copyright © 1984 by Van Nostrand Reinhold Company Inc.

Library of Congress Catalog Card Number: 84-3521

ISBN: 0-442-21737-4

Manufactured in the United States of America

Published by Van Nostrand Reinhold Company Inc.
135 West 50th Street
New York, New York 10020

Van Nostrand Reinhold Company Limited
Molly Millars Lane
Wokingham, Berkshire RG11 2PY, England

Van Nostrand Reinhold
480 Latrobe Street
Melbourne, Victoria 3000, Australia

Macmillan of Canada
Division of Gage Publishing Limited
164 Commander Boulevard
Agincourt, Ontario M1S 3C7, Canada

15   14   13   12   11   10   9   8   7   6   5   4   3   2   1

Library of Congress Cataloging in Publication Data

Coughlin, Vince.
   Telecommunications, equipment fundamentals and network structures.

   Includes index.
   1. Telecommunication.   I. Title.
TK5101.C694   1984   621.38   84-3521
ISBN 0-442-21737-4

TO SPECIAL AGENTS OF HAPPINESS:

My Mother, Nicholetta

Carol Ware Coughlin

Almira Figueira

PREFACE

Modern technology began in the 1950's and 1960's, with the development of transistor technology. At first it was useful in improving the performance of voice communications. But then it made possible extraordinary computer capability in manageable size—and at manageable cost. First came large mainframe computers for only the largest companies; and later the microcomputer as we know it today.

The increasing use of computers, in the 1960's with their ability to manipulate and store vast quantities of information, stimulated the need for computers to communicate with one another and so telephone circuits had to be segregated and conditioned specifically for computer traffic, using the modem. Computers ushered in a new era of business communications in which data could be developed, manipulated, stored or transmitted with remarkable ease.

The recent pace of technological advancement has been breathtaking and, today, the distinction between communications and computers is no longer even necessary. Computers, at the very core of communications networks, route and control communications on major common carriers.

The decade of the 1980's is bearing the fruits of the marriage of computers and communications. For the first time networks are enabling organizations to utilize the combined processing power of computers and communications equipment.

Hence both engineering and communications managers must have a thorough knowledge of telecommunications technology to meet the challenges of assembling such networks. These professionals will shape the electronic products and systems that will allow such activ-

ities as banking, education, travel, manufacturing and medical diagnostics to be performed more efficiently. By adding communications capability to these systems, users can benefit from having fast
and direct access to the enormous power of the computer.

The attempt of this book is to set in perspective the operating prin-
ciples that underlie the various telecommunications equipment and
network structures used to grow these systems. Having understood
the basics, the communications manager can make the proper technology choice for his applications. The book also serves as a user
guide to equipment that can fulfill requirements for setting up such
networks.

A special thank you to Debby Garcia for her assistance in typing
and preparation of this work.

Vince Coughlin

CONTENTS

Preface / vii

1. The Medium and the Market—An Overview / 1

  1.1 A Telecommunications Overview / 1
      1.1.1 Smart Telecom Banking Card Systems / 2
  1.2 Understanding the Basics / 3
      1.2.1 Parallel and Serial Transmission / 4
      1.2.2 Channels and Facilities / 4
      1.2.3 Digital Versus Analog Transmission / 6
      1.2.4 Asynchronous and Synchronous Transmission / 8
  1.3 A Bright Future with Fiber Optics / 9
      1.3.1 Fiber Selection / 10
      1.3.2 Light Source Emitters and Detectors / 12
  1.4 A View from the Business Side / 13
      1.4.1 How to Set your Requirements / 14
      1.4.2 Minimizing Line Costs / 14

2. Modems and Multiplexers / 16

  2.1 Introduction / 16
  2.2 The Role of Modems / 17
  2.3 Modem Applications / 19
  2.4 Industry Standards / 20
  2.5 Modem Features / 20
  2.6 Multiplexing Reduces Communications Cost / 22
  2.7 Multiplexer Technology / 24
  2.8 Statistical Multiplexer / 29
      2.8.1 Bandsplitter / 31

3. **Protocols and Codes / 32**

    3.1   What is a Protocol? / 32
    3.2   Protocol Hierarchy / 32
    3.3   Physical Electrical Interface / 34
        3.3.1   Handshaking / 35
    3.4   Link Control Structure / 37
        3.4.1   Data Transparency / 37
        3.4.2   Connection and Disconnection / 38
        3.4.3   Failure and Recovery / 38
        3.4.4   Error Control / 39
        3.4.5   Sequencing / 39
    3.5   Bisync Protocol / 39
    3.6   HDLC Protocol / 41
    3.7   SDLC Overview / 41
    3.8   System Network Architecture (SNA) / 42
    3.9   Polling Explained / 43
    3.10   Transmission Codes / 43
    3.11   Code and Speed Converters / 44
    3.12   Treatment of Errors / 45
        3.12.1 Redundancy Checks / 47
        3.12.2 Error-Correcting Codes / 47
        3.12.3 Loop Check / 48
        3.12.4 Effect of Errors / 48
    3.13   Front-End Processors / 49
        3.13.1 Selection and Evaluation / 49

4. **Terminal Technology / 51**

    4.1   CRT Display Terminals / 51
        4.1.1   Control Unit / 52
        4.1.2   Keyboard Structure / 53
        4.1.3   Display Schemes / 54
        4.1.4   Terminal Considerations / 54
    4.2   Printing Technologies / 56
    4.3   Graphic Terminals / 62
    4.4   Facsimile Technology / 63
        4.4.1   Stored Switching System / 66
        4.4.2   Image Resolution / 66
    4.5   Integrated Workstations / 67

5. **Network Management / 68**

    5.1 Network Control Design Considerations / 68
    5.2 Elementary Monitoring / 69
    5.3 Redundancy and Compatability / 70
    5.4 Network Testing / 71
    5.5 Centralized Troubleshooting / 71
    5.6 Network Control Center / 73
        5.6.1 Alarms / 73
        5.6.2 Automatic Trunk Testing / 74
        5.6.3 Maintenance and Management Reports / 74
    5.7 Specialized Test Equipment / 75
        5.7.1 Line Analyzers / 76
        5.7.2 Breakout Boxes / 77
        5.7.3 Error Rate Testers / 78
        5.7.4 Data Monitors / 79
    5.8 Network Security / 79

6. **Network Structures / 82**

    6.1 Topology Tradeoffs / 82
        6.1.1 Front-End and Back-End Networks / 84
    6.2 Local Area Networks / 86
        6.2.1 The Ethernet Way / 87
        6.2.2 LAN Cable Medium Choices / 89
    6.3 Distributed Communications / 91
    6.4 Message Switching Networks / 93
    6.5 Integrated Voice and Data / 95
    6.6 Shared Resources / 96
    6.7 Packet Switching Networks / 98
        6.7.1 The History of X.25 / 99
        6.7.2 How They Work / 100
        6.7.3 Packet Switching in Operation / 102
    6.8 Network Design Considerations / 103

7. **Satellite and Carrier Services / 107**

    7.1 Satellite Overview / 107
        7.1.1 Overcoming Attenuation / 109
        7.1.2 Satellite Communications Structure / 110

7.2   Selecting a Transmission Method / 111
    7.2.1   Private Alternatives / 113
    7.2.2   Measured Use Services / 114
7.3   Facsimile Services / 115

Glossary of Terms / 119

Index / 129

1
THE MEDIUM AND THE MARKET—
AN OVERVIEW

1.1 A TELECOMMUNICATIONS OVERVIEW

The many uses of telecommunications will change work patterns, leisure time, education, health care and industry.

Communications satellites will generate their own energy from sunlight in space. New optical-fiber cables and semiconductor lasers will work with them to replace copper wire. Such satellites and fibers could transmit all the information the human race could possibly use. Whatever the limits to growth in other fields, there are no limits near in telecommunications and electronic technology.

Imagine a home ten or twenty years in the future, with television sets which can pick up international channels, and also be used in conjunction with small keyboards to provide a multitude of communication services. The home has cabling under the streets and new forms of radio provide all manner of communication facilities.

Restaurants and stores all accept bank cards, which are read by machines that automatically transfer funds between bank accounts by telecommunications. Citizens can wear radio devices for automatically calling police or ambulances if they wish. Homes have burglar and fire alarms connected to the police and fire stations.

To avoid unemployment, long weekends have become normal and are demanded by the labor unions. Paperwork is largely avoided by having computers send orders and invoices directly to other computers and by making most financial settlements, including salary

1

payments, by automatic transmission of funds into appropriate bank accounts. What was once called the office-of-the-future has become as conventional as vending machines, and with plasma screens in briefcase lids, people take their office-of-the-future home with them.

1.1.1 Smart Telecom Banking Card Systems

Microchips are being placed in plastic cards that we carry in wallets, purses, and shirt pockets. "Smart" cards (cards containing microchips that can compute as well as hold data) could open a new frontier to designers of information systems, distributing processing power directly into the hands of the general public.

The smart card is made of a piece of plastic, incorporating an integrated-circuit chip with memory and computational capabilities. The memory is nonvolatile so it does not lose its contents when power is shut off.

The bank issues the card to an account holder, personalizing it with that individual's account number and choice of personal identification number (PIN) and inserting a secret code word associated with the bank. The bank also sets a monthly spending limit for the card. Only a bank official armed with the bank's code word can make or change an authorization, and only someone knowing the card holder's PIN can draw against it.

At the point of sale, the card is plugged into a terminal that has a separate customer keypad to protect privacy. The sales amount entered by the sales clerk through the main keyboard is displayed on the customer's keypad, and if it is acceptable, the customer enters his PIN. If this PIN agrees with the one stored inside the card and if the store's terminal has previously been activated by a legitimate store-owner's card, the transaction can proceed.

The terminal determines whether the purchase amount, plus the sum of the past month's purchases read from the customer's card, exceed the monthly authorization on the card. If there is enough credit left, the transaction is completed, and the transaction date and amount is written into the customer's card.

At the same time the date, amount, and the purchaser's account number are entered into an electronic memory inside the store's terminal. In this "store and forward" terminal attachment mode, an entire transactions file is delivered—daily or at any convenient interval—to the store's bank via a dialed telephone line or by physically transporting the memory module to the bank. The bank clears the transactions by the electronic funds transfer (EFT) from the purchaser's account to the store's account.

This transactional system performs virtually the same functions as automatic teller machines (ATMs) and self-service travelers' check dispensers, but it differs fundamentally from them by using the computational capability and memory of the smart card in place of the on-line capabilities of a host computer.

The smart card could "personalize" communications terminals, such as home terminals and telephones, acting both as the payment vehicle and the record keeper. It would be very helpful in the medical-services field, using its computational power to authenticate claims and carry personal medical histories and treatment records between interdependent, but separate, medical-information systems. It could become a social-services eligibility card, reducing the costs and errors of paper-based systems. It could replace identification documents, such as passes to secured areas, passports, and alien identification cards.

1.2 UNDERSTANDING THE BASICS

In order to understand current trends in telecommunications and to appreciate the direction of future trends, it is necessary to understand the basics.

Let us first look at the elements of a basic telecom network. A network is an arrangement of transmission, switching, signaling and terminal equipment. The transmission system provides the electrical path for information to flow from one location to another. Switching includes identifying and connecting independent transmission links to form a continuous path from one location to another. Signaling involves supplying and interpreting the supervisory and address signals needed to perform the switching operation.

Many times organizations with many branches or regional facilities have the need to transmit and receive data frequently. They may use a data communications system consisting of a wide variety of components and many times a variety of services. The combination chosen will depend upon the merits and limitations of various networks. Perhaps one or more host processors may go through multiple front ends into one or more types of networks involving various types of terminals. A network of this type is called a "distributed network."

The electronic transmission of encoded information from one point to another requires various physical elements, devices, and systems, as well as standards and procedures. An understanding of these basic elements and concepts can help users of telcom services to take advantage of the communication systems that are now available.

1.2.1 Parallel and Serial Transmission

Short distance communications between terminals (data input/output devices) and computers may take place in a variety of ways. Some configurations, for example, involve the transmission of data characters across multiple lines. This is known as parallel transmission. If you were transmitting one bit of a seven-bit ASCII code character per line, you would need seven lines. This method of transmission is called parallel-by-bit serial-by-character (see Fig. 1-1). This method is often used over short distances. However, over long distances (involving miles) serial transmission is preferred because it involves only one data line.

In order to send data over a single line, the bits must be transmitted in series. This is called (appropriately) serial transmission. It is the most widely used method of transmitting data. In seven-bit ASCII, a "C" would appear as illustrated in Fig. 1-2.

1.2.2 Channels and Facilities

A communications link or channel is a path for electrical transmission between two or more stations or terminals. It may be a single wire, a group of wires, a coaxial cable, or a special part of the radio frequency spectrum. The purpose of a channel is to carry informa-

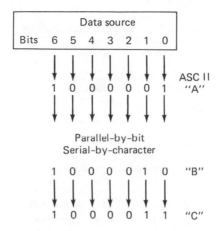

Fig. 1–1. Parallel-by-bit and serial-by-character transmission.

tion from one location to another. All channels have limitations on their information handling abilities, depending upon their electrical and physical characteristics.

There are three basic types of channels: simplex, half-duplex and full-duplex. As an example of each, consider transmission between points A and B in Fig. 1–3. Transmission from A to B only (and not from B to A) requires a simplex channel. Simplex channels are used in loop mode configurations such as supermarket checkout terminals. Transmission from A to B and then from B to A, but not simultaneously, requires a half-duplex channel. If a two-wire circuit is used, the line must be turned around to reverse the direction of transmission. A four-wire circuit eliminates line turnaround. Transmission from A to B and from B to A simultaneously describes a full-duplex channel. Although four wires are most often used, a two-wire circuit can support full-duplex communications if the frequency spectrum is subdivided into receive and transmit channels.

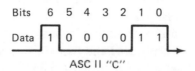

Fig. 1–2. Serial transmission is most often used.

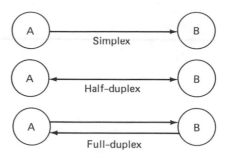

Fig. 1-3. Types of communications channels.

In addition to the direction of transmission, a channel is characterized by its bandwidth. In general, the greater the bandwidth of the assigned channel, the higher the possible transmission speed. This speed it usually measured in terms of the number of line signal elements per second, the baud rate. If a signal element represents one of two binary states, the baud rate is equal to the bit rate. When more than two states are represented, as in multilevel modulation, the bit rate exceeds the baud rate.

1.2.3 Digital Versus Analog Transmission

Digital transmission can be applied to digital data or analog voice signals. In either case, information is sent over the communications channel as a stream of pulses. If analog to digital conversion is required, the signal voltage values are sampled and represented in binary format (see Fig. 1-4). Pulses transmitted over a communication line are distorted by line capacitance, inductance, and leakage. The longer the line or the faster the pulse rate, the more difficult it is to interpret the received signal. This signal degradation is the reason for the closely spaced regenerative repeaters used in digital data transmission facilities. When noise and distortion threaten to destroy the integrity of the pulse stream, the pulses are detected and regenerated. If the regeneration process is repeated properly, the received signal will be an exact replica of the transmitted signal. It is possible to transmit pulses over short distances using privately owned cable or wire pairs. This is baseband transmission and usually requires line

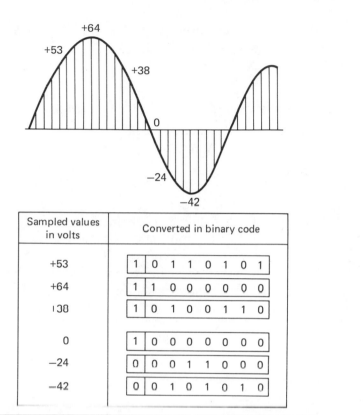

Fig. 1-4. In a digital network analog signals must be converted into digital bits.

drivers and receivers on each end of the line. Longer distance communication must use the digital transmission facilities of the common carriers.

In analog transmission, a continuous range of signal amplitudes or frequencies is sent over the communications line. Linear amplifiers maintain signal quality. The voice telephone network supplied by the common carriers uses analog transmission facilities to service most data communications users. To interface the analog voice channels to digital terminals and computers, a modulator-demodulator (modem) is used. In a modem, digital information modulates a carrier signal, which passes through the telephone network just as does a voice signal. At the receiving end, the signal is demodulated back into digital form.

1.2.4 Asynchronous and Synchronous Transmission

Asynchronous data are typically produced by low-speed terminals with rates of less than 1200 baud. In asynchronous systems (Fig. 1–5a), the transmission line is in a mark (binary 1) condition in its idle state. As each character is transmitted, it is preceded by a start bit, or transition from mark to space (binary 0), which indicates to the receiving terminal that a character is being transmitted. The receiving device detects the start bit and the data bits that make up the character. At the end of the character transmission, the line is returned to a mark condition by one or more stop bits, and is ready for the next character. (An asynchronous character varies in length depending on the information code employed.) This process is repeated

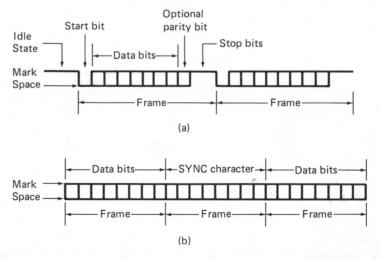

Fig. 1–5. Asynchronous and synchronous transmission. Choice of proper timing mode depends directly on application. Asynchronous transmission (a) is used mostly with man-machine interfaces; synchronous transmission (b) offers high speed necessary for machine-machine communication.

character by character until the entire message has been sent. The start and stop bits permit the receiving terminal to synchronize itself to the transmitter on a character by character basis.

Synchronous transmission (Fig. 1–5b) uses an internal clocking source within the modem to synchronize the transmitter and receiver. Once a synchronization character (SYN) has been sensed by the receiving terminal, data transmission proceeds character by character without the intervening start and stop bits. The incoming stream of data bits is interpreted on the basis of the receive clock signal supplied by the modem. The receiving device accepts data from the modem until it detects a special ending character or a character terminal count at which time it knows that the message is over. The message block usually consists of one or two synchronization characters, a number of data and control characters (typically 100 to 10,000), a terminating character, and one or two error control characters. Between messages, the communication line may idle in SYN characters or be held to mark. Note that synchronous modems can be used to transmit asynchronous data, and, conversely, asynchronous modems can be used for synchronous data if the receiving terminal can derive the clock signal from the data.

Asynchronous transmission is advantageous when transmission is irregular (e.g., when it is initiated by a keyboard operator's typing speed). It is also inexpensive because of the simple interface logic and circuitry required. Synchronous transmission, on the other hand, makes far better use of the transmission facility by eliminating the start and stop bits on each character. Furthermore, synchronous data are suitable for multilevel modulation, which combines two or four bits in one signal element (baud). This can facilitate data rates of 4.8 k or 9.6 k bits per second over a bandwidth of 2.4 kHz.

1.3 A BRIGHT FUTURE WITH FIBER OPTICS

A light-bearing finger-sized fiber-optic cable can carry 40,000 phone calls at once in contrast to 20,000 calls with an arm sized copper cable. The conduit for sending these super-quick light pulses is made of glass, with clarity 1000 times greater than ordinary window glass.

Table 1-1. Fiber Optics versus Conventional Cable

	SOLID COAX	FLAT RIBBON	CATV COAX	FIBER OPTIC
Physical size	Small	Medium	Large	Small
Mechanical integrity	Good	Excellent	Excellent	Fair
Weight	Low	Moderate	Heavy	Low
Crosstalk immunity	Good	Good	Good	Excellent
Bandwidth	Good	Fair	Good	Excellent
EMI and noise immunity	Good	Fair	Good	Excellent

Fiber optics systems offer an effective alternative method to transmit information (see Table 1-1). Fiber optics provide the advantages of a light weight, smaller size, durable, nonconducting signal path which has no effect on the electrical environment that the signal passes through. In addition, the frequency and distance limits of a fiber optic system are said to be better than twisted pair, coaxial cable, or microwave waveguide systems.

Fiber optic systems (Fig. 1-6) have three major components: the light sources in the transmitter, the fiber cable, and the photodetectors in the receiver. The sources must be inexpensive, have long lifetimes, and be easy to modulate at high frequencies. The fiber must be sufficiently transparent so that the signal does not have to be refreshed, amplified and repeated for short distances. The photodetectors must be inexpensive, sensitive to the proper frequencies, and exhibit low noise levels.

1.3.1 Fiber Selection

A choice of fiber type determines the amount of power the emitter or light source can place in the fiber and the attenuation of the light

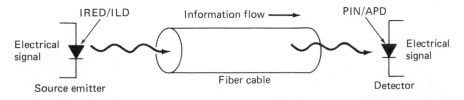

Fig. 1-6. Major elements of a fiber optics systems.

power as it travels down the fiber. A basic understanding of this can be obtained by comparing various fiber types. A step index fiber has an abrupt change in the index of refraction. In a graded index fiber, the index of refraction changes gradually from a high to a lower value as position varies radially from the center. This causes the rays to gradually bend back towards the center, instead of reflecting, and keeps the rays traveling along the fiber at the same velocity. This velocity matching property in a multimode fiber generally gives graded index fibers higher bandwidth-distance characteristics. The highest bandwidth-distance fibers are designed to operate like microwave waveguides. (see Fig. 1-7).

The velocity of light in the fiber, as in any medium, varies with its wavelength—the longer the wavelength, the greater the velocity. Thus, if the source emits light over a range of wavelengths, the shorter wavelengths of light will trail behind the longer, broadening the signal. However, if the diameter of a fiber is less than about two wavelengths, only one mode will be propagated, which is called a single mode fiber. At this diameter, fiber optics begins to resemble the transmission of microwaves.

Single mode fibers provide a greater bandwidth than do multimode fibers because the signal is not subjected to modal dispersion. However, less power can be coupled because of the severe constraint imposed by the single characteristic angle. The development of sin-

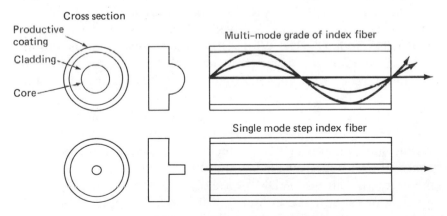

Fig. 1-7. Signal propagation modes for fiber types.

gle-mode fiber through the 1980's is expected to encourage new applications for fiber optic technology and to allow transmission of signals greater distances without repeaters.

1.3.2 Light Source Emitters and Detectors

The emitter transforms an electrical signal into a light signal which can be coupled to the fiber and transmitted. The two most popular light sources used are infrared emitting diodes (IRED) and injection laser diodes (ILD).

The laser diodes provide a small area, powerful source with high speed capability. They are exceptionally well suited to digital transmission systems. They are highly directional emitters and substantially more power may be coupled if the injection laser diode is used.

Light emitting diodes are ideally suited to analog applications. However, their numerical aperture is high, leading to a large mismatch when coupled into fibers with small cones of acceptance. It is therefore often difficult to achieve good power couplings from most LED sources into small diameter fibers.

Fiber optic receivers, or detectors, are photodiodes which convert light to electrical current. Photodiodes are used because they are small, inexpensive, and do not need much power to drive them. There are two types of photodiodes in common use today: the PIN diode and the avalanche diode, or APD. The PIN diode is the most simple and inexpensive type of photodiode, and is adequate for a wide range of applications. A more sensitive, and more expensive, type is the avalanche diode (APD). With APD's the internal gain is amplified to produce a larger current than the PIN diode. Also, noise causes less interference with an APD than with a PIN diode, which allows for higher transmission speeds along the link.

The commercial availability of light sources and detectors operating in the longer wavelength regions, together with wavelength division multiplexing (WDM) techniques will find fiber optics competing with microwave for high-bandwidth, long-distance transmission applications. WDM especially will come into play as communications suppliers continue to lean towards supplying broadband services to subscribers, such as video and data. With WDM, multiple optical signals may be transmitted over different wavelengths within a single optical fiber.

WDM is accomplished by optically combining outputs of several light sources, each operating at a different wavelength or color, and coupling the composite signal into an optical fiber. An optical splitter, or demultiplexer, at the receiving end of the link separates the various wavelengths or colors and channels them into separate receivers. Because the different color light signals travel through the fiber independently, each represents a high-bandwidth transmission channel. WDM is expected to be used on a regular commercial basis by the mid-1980's.

1.4 A VIEW FROM THE BUSINESS SIDE

The telecommunications market is an eight billion dollar market with expected growth to thirteen billion dollars by 1985. The protion of the communication market composed of large (Fortune 500/50) companies represents 40% of the total business communications market. This large company market is changing at a higher rate than most other major segments of the communication market. These changes include high growth in some segments of the market as well as high levels of equipment upgrades in other segments. Within this market, significant business opportunities are created.

The primary key to the market is providing user controls. The driving force behind most of the network changes is the users' desire for control. Control includes network management, maintenance, and user access control. A major result of these changes is that the fastest-selling products in the market incorporate network control capabilities. These include call detail recording systems in voice networks and centralized network control systems in data networks.

Cost reduction is a very common driving force in communication networks, particularly in the voice and message area. Technology impacts the users' network changes in terms of allowing their requirements to be met.

1.4.1 How to Set your Requirements

Like the planners of expressways in urban areas, planners of tele-communication systems sometimes find that the best way to do things in not always a permissible way. A systems planner must always work within the constraints of company policy, number of personnel available, security restrictions, and budgetary considerations.

Be careful to evaluate the time savings associated with transmission for each set of data. If the processing cycle, once data is received, is much longer than the time to transmit it, little may be saved by using data communications. For example, say the processing cycle requires four days once data has been received. If the alternative to transmission over a communications line is next-day mail delivery, very little has been saved in the overall cycle. This question is sometimes complicated by multiple use of transmitted data. Sales data from a branch office can be used to generate shipping orders, invoices, inventory replenishments and sales analyses, each with a different time priority.

This leads to a second consideration—extent of data collected. In the example just cited, extra information is required for each intended usage. And each little bit adds to the transmission time, the processing time, etc.—all the way down the line. This is not to say that the additional uses should be eliminated and only the bare bones transmitted. After all, the incremental cost to handle a little bit more is usually small in comparison to the total cost of the system. But even a small incremental cost compounded many times can add up to a sizable cost once in motion.

1.4.2 Minimizing Line Costs

As the size and volume of the transmitted messages grow, a closer look is taken at the communications lines connecting the terminals with the computer. Because the cost of the communications lines can easily amount to one-third of the cost of the whole system, economies here can have important budgetary effects.

Several techniques are available to reduce the line costs and perhaps improve performance at the same time. A private line can be leased from the telephone company, giving exclusive use to the customer for 24 hours a day. If the volume from one location is insufficient to warrant a private line, perhaps several locations can share the same line in a multi-drop fashion. This requires that the terminals be equipped to recognize their own addresses, so that information directed to one terminal won't be sent to all the other terminals. It also introduces the problem of multiple stations competing for use of the one line. With the dial-up telephone network, the telephone company equipment takes care of this problem for you automatically. Another capability that is achievable on a multi-drop line is broadcasting data to a selected group of terminals simultaneously.

You may also improve the cost-effectiveness of your data communications system through alterations in the connecting links and by examining a variety of other services available from the telephone company, specialized carriers, and private companies.

2
MODEMS AND MULTIPLEXERS

2.1 INTRODUCTION

Modems and multiplexers are essential components of data communications networks. Both are needed to support communications between a central site processor and remote terminals or network nodes. Modems are necessary to support digital communications over long distances. They convert digital signals from data terminal equipment to an analog form for transmission over analog communications facilities, and reconvert the analog signal to a digital form at the other end. Multiplexers combine transmissions from several sources on one line, eliminating the cost of several lines and modems between two points to support independent transmissions. Modems and multiplexers are available from a multitude of vendors. The products offered by these vendors differ in application, performance, and price. Selecting the appropriate modems and multiplexers to satisfy user requirements is often a compromising task requiring considerable time and effort. Network managers and planners need a sound understanding of data communications technology and an awareness of the available products and their specifications to design a cost-effective network using modems and multiplexers that meet current and future needs. This chapter provides a basic understanding of modem and multiplexer technology, explains essential operating parameters, and describes general application areas.

2.2 THE ROLE OF MODEMS

A modem converts a digital bit stream (Fig. 2–1) from a computer or terminal to an analog signal which it transmits over a communications medium, typically a telephone line. A modem at the other end of the line converts the analog signal back to the digital bit stream used by the attached data terminal equipment. Digital to analog conversion is necessary to establish communications compatibility with the electrical characteristics of a telephone line, which are suitable to voice communications, but hostile to digital transmission.

The term modem is an acronym for modulation/demodulation. Digital signals are converted to analog form through a process called modulation, which uses the digital signal to alter a fixed signal called the carrier. The modulated carrier signal contains the digital data, which is extracted from the carrier by the receiving modem through a process called demodulation. Simple modulation techniques are used for low data rates, but as the data rate is increased, more complex methods are required.

Frequency shift keying (FSK) is the simplest form of modulation used by low speed modems. This commonly used frequency modulation technique transmits digital data via two unique frequencies or tones (Fig. 2–2), one assigned to binary ones (marks), the other assigned to zeros (spaces). Phase shift keying (PSK) is a phase modulation technique commonly used by medium and high speed

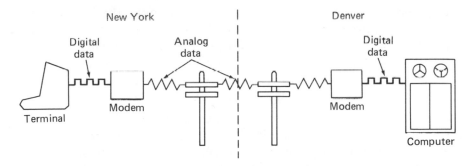

Fig. 2–1. Role of the modem.

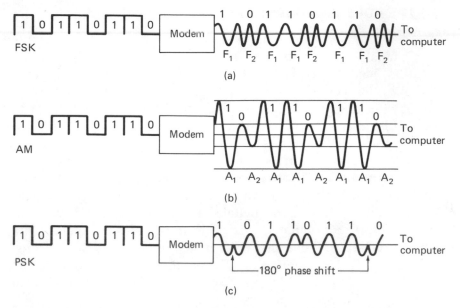

Fig. 2-2. Modulation techniques. In frequency shift keying (a), digital signal (left) received by modem is shifted in frequency to represent binary 1 and binary 0. In amplitude modulation (b), one carrier signal amplitude (A1) represents binary 1, and another (A2), binary 0. In phase modulation (c), whenever 180° phase shift is encountered, receiving modem assigns value of binary 0 and at all other times, value of binary 1 is assumed.

modems. The phase of the modem's carrier signal is shifted by a fixed multiple with respect to a known reference (clock frequency) for assigned digital bit combinations. For example, the Bell System 201 modems use a four-phase PSK technique to achieve a data rate of 2400 bps. The four combinations of bit pairs (dibits) are encoded as 90-degree phase shifts.

Similarly, three bits (tribits) can be transmitted with each 45-degree phase shift using an eight-phase PSK technique, to achieve a data rate of 4800 bps. Higher data rates require even more sophisticated and complex modulation techniques to compensate for the limitation imposed by voiceband facilities. Quadrature amplitude modulation (QAM) is commonly used to achieve data rates of 9600 bps or higher within a 3 kHz bandwidth. QAM modulation combines amplitude and phase modulation to encode multiple data bits with each signal change.

Multiple bit encoding techniques yield a higher data rate in bits per second (bps) than the actual signaling rate in bauds. Baud is an old telephone term defined as changes per second. Baud rate is equivalent to bits per second only when a signal change represents a single bit; baud rate is not equivalent to bits per second with multiple bit encoding, where each signal change represents multiple bits.

2.3 MODEM APPLICATIONS

Modem vendors produce modems for various applications. Each is designed to meet specific operating criteria. Application categories include dedicated point-to-point or multipoint communication networks, the public switched telephone network (DDD), and private facilities. Modems designed for dedicated facilities fall into two categories: voiceband (or voice-grade) and wideband. Voiceband modems communicate at rates up to 19.2 k bps with a 3 kHz bandwidth, such as type 3002 service provided by the telephone company. Wideband modems communicate at rates of 19.2 k bps and above and require broadband facilities such as the Bell System series 8000 wideband service or equivalent.

Modems designed for multipoint communications reduce the training time required for modem synchronization as compared with point-to-point modems. Fast training time significantly reduces the time delays, which can be substantial when there are many drops per line.

A special type of modem called an acoustic coupler acoustically connects the modem to the DDD network. The coupler employs FSK modulation and couples acoustically with a telephone handset. Acoustic couplers are useful for portable applications, and they are typically priced below conventional modems. However, they are limited to transmission speeds of 1200 baud.

Modems designed for private facilities are called limited distance or short-haul modems. Private facilities are typically those provided by the user for communication within a building or between buildings in the same complex. These modems are less expensive than long-haul modems and eliminate the cost of lines provided by the telephone company. The transmission range is usually limited to 4 or 5 miles.

2.4 INDUSTRY STANDARDS

Data communications standards have been established by several industry standards organizations. The Electronic Industries Association (EIA) and the Consultative Committee for International Telegraph and Telephone (CCITT) have established standards for modem parameters and the electrical interface that provides a connection between data terminal equipment (DTE) and data circuit terminating equipment (DCE); such as a modem (see Fig. 2–3). The interface passes digital data and control signals between the devices, but can differ electrically depending on application.

The EIA Standard RS-232C is the most common electrical interface standard used throughout the industry; however, some connections support functions that differ among vendor products. The physical RS-232C interface is a 25-pin connector that supports transmission at data rates up to 20 k bps at distances to 50 feet between DTE and DCE. EIA RS-449 is an improved interface standard compatible with RS-232C, but it accommodates extended data rates to 2 m bps at distances up to 200 feet between DTE and DCE. The added connections of the 37-pin interface extend control functions. The U.S. Government requires the RS-449 interface on all data communications products that it acquires. Industry has been sluggish about adopting the improved standard for commercial applications.

2.5 MODEM FEATURES

Modems are available with several useful features that benefit specific applications. Most high-speed modems are available with a multiport feature, to accommodate two or more (typically four) terminals or lines with aggregate data rates that do not exceed the modem's maximum data rate.

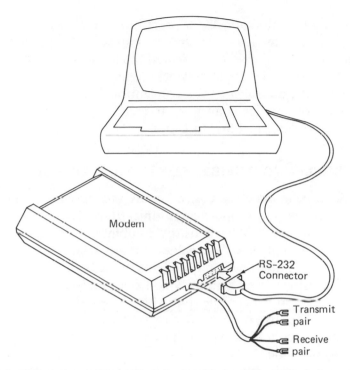

Fig. 2-3. Modems are supplied with a simple cable terminal with spade lugs, which are either two- or four-wire depending on network type. The telephone company provides a terminal block or an 829 four-wire terminating set. The interface to the terminal is RS-232 or RS-449.

A reverse channel is an economical alternative to full-duplex communications on a four-wire line to support message acknowledgement. The narrowband simplex channel eliminates turnaround time for message acknowledgement with the economy of a two-wire line.

A secondary channel is a full-duplex narrowband channel that typically supports data rates up to 150 bps and shares the bandwidth of a four-wire line with the main channel. The out-of-band channel is useful for several applications. For example, it can be used as a data channel for a low-speed terminal to eliminate the cost of another line. Alternate voice data eliminates extra charges for voice-coordinated transmissions. The feature includes a telephone handset and dialer with switch selection of voice or data.

Auto answer automatically answers incoming calls over the DDD network for unattended sites. Dial backup restores communications interrupted by a failure on a dedicated facility by reestablishing communications over the DDD network. The dial-up connection is established at the central site via a conventional telephone; remote site equipmemt automatically answers an incoming call and switches the modem from dedicated line to DDD.

2.6 MULTIPLEXING REDUCES COMMUNICATIONS COSTS

Multiplexers eliminate the cost of many individual communication lines and modems to support communications between central computer sites and remote terminals, by combining the individual transmissions on a single line. The value of multiplexing is realized when the cost of separate lines and modems exceeds the cost of a single line, a pair of modems, and back-to-back multiplexers.

Multiplexers are adaptable to a wide variety of configurations to achieve the objective of reduced operating costs. Equally important, multiplexers accommodate growth in traffic volume without the need for additional lines. Large scale multilink multiplexers link nodes within a multinode network. This arrangement benefits the exchange of data between the different nodes, which may be located in different cities, and provides multiple data paths to reroute data in the event of a link outage. Multiplexers can also accommodate low-speed asynchronous terminals with different transmission rates over DDD lines, eliminating the cost of separate lines and ports to support the different transmission rates.

Multiplexers are available in a wide range of channel capacities to satisfy small, medium, and large scale network requirements. Models are available to support as few as four or eight terminals and/or lines, while top-of-the-line models are available with more than 200 ports. Medium and large scale units are easily expanded to meet growth requirements or reconfiguration needs. Most state-of-the-art multiplexers are available with an integral composite link modem. This eliminates the added cost of a separate modem, saves space, and puts multiplexer and modem under the same service umbrella.

A typical multiplexer system with a computer located in New York

City and eight remote terminals in Denver is shown in Fig. 2–4. The multiplexers allow eight terminals in Denver to share one telephone line where previously eight lines were required. It should be noted that multiplexers are normally required at both ends of a shared telephone line, thus allowing the eight channels multiplexed at one line to be demultiplexed at the other, thus providing a transparent connection between each of the computer ports to each of the remote terminals. The word transparent means that the multiplexer does not interrupt the flow of data and EIA RS-232 interface signals, which

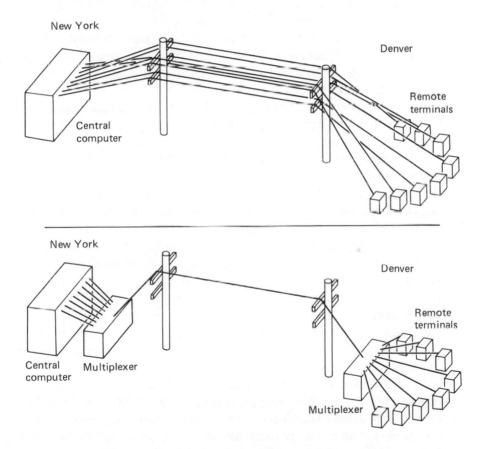

Fig. 2–4. A typical multiplexer system.

normally pass through the cable between the computer port and the attached modem or terminal.

When multiplexers are installed, neither the computer nor the modems located in Denver know the multiplexer system is being used, whether the telephone lines attaching the terminals to the mulitplexer are leased lines or dial-up lines. As a result, neither terminal equipment nor computer software need to be changed when multiplexer systems are installed to reduce telephone line rental or long-distance dial charges.

2.7 MULTIPLEXER TECHNOLOGY

Multiplexers combine multiple communications paths by one of two methods: frequency division or time division. Each technique combines multiple paths on a single link in a different manner to satisfy diverse needs.

Frequency division multiplexing (FDM) is the earliest and least sophisticated form of multiplexing. Applicable to low-speed telegraph and data transmissions as well as voice, FDM divides the bandwidth of a voiceband line into multiple frequency bands or derived channels. A narrow band of frequencies called a guard band is required to separate adjacent channels to prevent signal interference between channels. Data transmission is FSK modulated. FSK (frequency shift keying) employs a pair of frequencies to correspond to a binary one and binary zero. The sending and receiving multiplexer channels can use the same set of frequencies for full-duplex operation over four-wire lines. Full-duplex operation on two-wire lines requires a different pair of frequencies for sending and receiving.

The maximum number of channels that can be supported on a voice-grade channel using the FDM technique is limited by the data rate of each channel. Bandwidth is proportional to data rate; higher data rates require greater bandwidth, reducing the available bandwidth for additional channels. The guard bands between channels further reduce the available bandwidth. The FDM technique is, therefore, a practical solution to combining transmissions from low-speed devices such as teleprinters that do not exceed 1200 baud. The FDM technique (Fig. 2–5) offers several advantages. No modem is

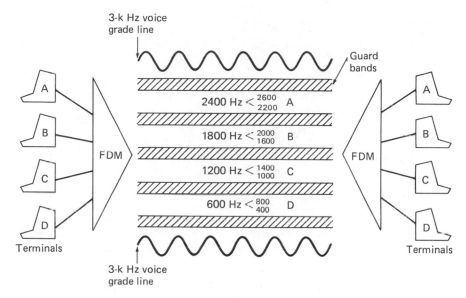

Fig. 2–5. With frequency division multiplexing, division depends on desired frequency of transmission, e.g., 75-Baud channel requires less spectrum space than a 50-baud channel.

required for the composite link since the multiplexer composite, or aggregate, consists of modulated analog signals, not digital bit streams. Full-duplex operation can be achieved on a two-wire line although only half the bandwidth of a four-wire line is available. Individual channels can be dropped at remote sites to support a single device. Voice and data can be mixed and the FDM technique is the most economical.

Time-division multiplexing (Fig. 2–6) is a digital technique. It interleaves bits (bit TDM) or characters (character TDM), one from each attached channel, and transmits them at high-speed down a telephone line equipped with high-speed synchronous modems. At the other end, another multiplexer demultiplexes the "bit train," or "character frame," presenting one bit or character to each low-speed channel just as they originated.

The statistical-multiplexer, sometimes referred to as a data concentrator, uses a relatively new technique. Like the TDM it is digital but utilizes microcomputer technology to provide greater efficiency and automatic retransmission.

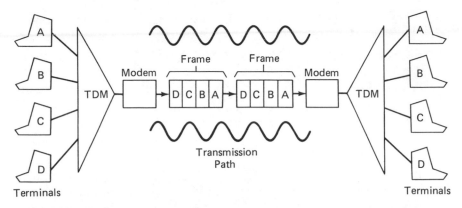

Fig. 2–6. Time division multiplexing. Time-slots are allotted to channels on dedicated basis whether or not "live" data are being carried. At any given time, majority of slots may not carry useful data but contain fill characters.

These three multiplexing techniques are compared on the basis of general criteria as shown in Table 2–1. Frequency division multiplexing (FDM) does not score well because it is not a digital technique. It is inefficient because the channels have to be separated across the band to prevent cross-talk and this, or course, wastes bandwidth. FDM is inflexible because a change in channel speed or number of channels may demand that the center frequencies of all channels be redefined. Channel capacity is also physically limited by the telephone channels 3000-Hz bandwidth resulting in a capacity of only six channels at 300 bps. Since the FDM is not capable of multiplexing two channels at 1200 bps, it scores very poorly as a device for handling high speed channels.

One strong point of the FDM, however, is that since each channel

Table 2–1. Multiplexing Technique Comparisons

COMPARISON FACTOR	FDM	BIT TDM	CHARACTER TDM	STAT MUX
Efficiency	Poor	Good	Excellent	Excellent
Channel capacity	Poor	Good	Good	Excellent
Flexibility	Poor	Good	Good	Excellent
Reliability	Good	Excellent	Excellent	Good

uses a different channel within the bandwidth, individual channels can be dropped at different points along the same telephone line. This allows multiplexing of one or two terminals in one city or location and extending the line to the next city or location to pick up one or two more terminals. There can be as many drops along the line as there are channels along the capacity of the FDM system. Thus the FDM provides an attractive multidrop capability for simple low-speed asynchronous terminals without imposing any requirement for addressing or special communication protocols.

Since the TDM is a digital system, it can be used with more complex synchronous modems capable of transmitting data at speeds up to 9600 bps on voice-grade telephone lines. It may also be used at much higher speeds on the newer DDS (dataphone digital service) circuits or satellite channels operating at speeds of 56,000 bps or higher. TDM's allow considerable flexibility in changing the number and/or the speed of channels as the requirements of the network change. This is possible because the TDM is a digital system independent of the data modem attaching it to the high-speed line. The small amount of buffering in the TDM allows it to regenerate data on each channel, which tends to reduce the overall error rate caused by end-to-end bit and character distortion.

As previously mentioned, there are two types of TDM's, the bit TDM and the character TDM. The character TDM is primarily used when multiplexing asynchronous channels. The character TDM buffers a complete character before transmitting it down the high-speed line. It removes the start and stop bits from the character before transmission, then adds the start and stop bits back to the character during the demultiplexing process at the other end. Thus, in the case of a dumb terminal application, it is only necessary to transmit eight bits of data for every character received. The net result is that the high-speed line is used more efficiently and channel capacity on a given high-speed line is greater with character TDM than with bit TDM.

The simplicity of the TDM technique results from the fact that the multiplexers communicate with each other, transmitting a constant stream of bits (bit TDM) or characters (character TDM) with a regularly recurring SYNC character in a constant number of "time-slots" between SYNC characters. Each time-slot contains a prede-fined number of bits or characters for a specific channel, as shown in Fig. 2–7a.

The fundamental principle of time-division multiplexing is that the frames (the data block following each SYNC character) are fixed in length and are transmitted continuously, since the receiving multi-plexer only knows which bits or characters belong to each channel as a function of their time relationship to the SYNC character. The number of bits or characters in each "time-slot" allows it to handle its assigned channel without losing data even when the channel is operating at maximum speed. But since the number of bits or char-acters is fixed, the TDM must transmit them even if the TDM is operating below its maximum speed or not operating at all. Since transmission is continuous, there is no possibility of retransmission even if the receiving multiplexer detects errors. The "idle" character used to fill a "time-slot," when no data character is ready for trans-mission, is also used to transmit channel interface condition infor-

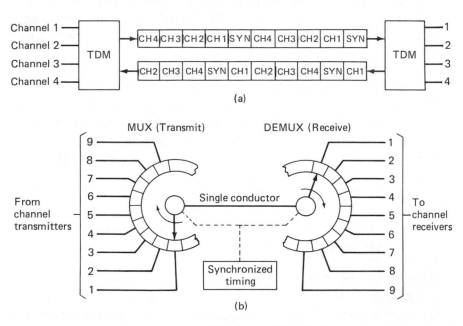

(a)

(b)

Fig. 2–7. TDM with fixed sequential scanning.

mation, remote loopback commands, and diagnostic status information.

Figure 2–7b shows that a TDM configuration with fixed sequential scanning samples a bit or character from each channel in order; channel 1 then 2, 3, etc. Thus all channels occupy the same amount of bandwidth on the high-speed line. If the highest speed terminal attached to the multiplexer is 300 bps, all channel allocations on the high-speed line will be at 300 bps regardless of the terminal speed, which may be lower. To counteract this deficiency, most TDMs on the market today use variable channel scanning.

A variable scan multiplexer has the ability to make three samples for each 300-bps terminal, compared to one for 110 bps or two for 134.5-bps terminals. This permits more efficient use for high-speed line capacity. But, with or without variable channel scanning, each channel's time-slot is fixed and capable of supporting the channel operating at maximum speed continuously.

Unfortunately for TDM efficiency, it is rare for any terminal to transmit continuously at its maximum operating speed. This is particularly true of the interacting dumb terminals used with minicomputer systems, which typically transmit only a line or perhaps a screen full of data before pausing.

2.8 STATISTICAL MULTIPLEXER

In a typical application, it is rare to find a terminal transmitting continuously at its maximum operating speed. The statistical multiplexer takes advantage of this and allows for a larger number of terminals to share a single phone line. Through the use of microcomputer technology it is designed to statistically determine traffic and transmit variable length blocks according to the loading on individual channels. In addition, it can check these transmitted blocks at the receive end for errors and automatically request retransmission in the event an error is detected. This is all accomplished without the need for any additional intelligence in the computer or terminal equipment in the network.

A comparison is shown in Fig. 2–8 between a TDM and a statistical multiplexer. As shown, the TDM, in order to operate four chan-

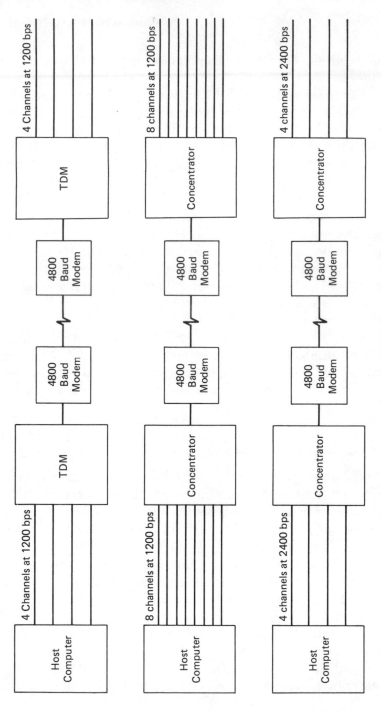

Fig. 2-8. Comparison between a TDM and a statistical multiplexer.

nels at 1200 baud, would require a 4800-baud high-speed link between the multiplexers. The statistical multiplexer on the other hand, may permit eight channels at 1200 baud to operate over the same 4800-baud data link, or even four channels at 2400 baud, operating over the 4800-baud data link. It is common in a statistical multiplexer application to have the aggregate speeds of the terminals to be equal to or even greater than the speed of the synchronous communication line between the multiplexers. This is permitted due to data buffering in each of the multiplexers.

2.8.1 Bandsplitter

Synchronous transmissions with error correction/detection can share the bandwidth of a composite link with the input/output of a statistical multiplexer via a bandsplitter. The bandsplitter is simply a bit-interleaved TDM that splits the bandwidth of the composite link into separate time-slots assigned to each of the synchronous channels and the statistical multiplexer. This arrangement minimizes delay time for the synchronous channels, but reduces bandwidth assigned to the statistical multiplexer. Since maximum data rate over a line is proportional to bandwidth, a reduction in bandwidth reduces the composite link data rate of the device. Applications with moderate to heavy channel loading will fill the multiplexer buffer quicker and more frequently, engaging flow control more frequently to interrupt transmission, resulting in performance degradation. Furthermore, time-slots assigned to the synchronous bandsplitter channels are empty during periods of inactivity, which also reduces communications efficiency. Bandsplitters are useful only when the traffic loading of the statistical multiplexer is not seriously effected.

3
PROTOCOLS AND CODES

3.1 WHAT IS A PROTOCOL?

Just as protocols are needed for an orderly society, data networks require their own set of rules to ensure the smooth exchange of information between communicating equipments. As data networks grow in complexity, so too will software provided communications protocols take on added importance. Until now, most of the attention surrounding protocols has focused on the so-called line control procedures, or data link protocols, which provide the rules by which two or more machines may converse, or exchange information, over a data link in an efficient and reliable manner. Minimally, the protocol must offer a means of identifying the sender and receiver on a multipoint or dial-up network, plus a way to indicate the start and end of the information being transmitted, and a method for detecting errors and initiating a corrective action.

To a degree, link control procedures will continue to be the most important kind of protocol in the years ahead, since they are so basic to the functioning of data networks. However, such procedures represent only one layer of a hierarchy of protocols, including ones to ensure tramsmission integrity over mutiple links, or through multiple networks, and others to allow processes within one computer to communicate with processes in other computers.

3.2 PROTOCOL HIERARCHY

A protocol hierarchy has been systematized in a reference model created by the International Organization for Standardization (ISO) to guide standards efforts involving computers and communications. The multi-level approach has a number of advantages. First, the protocol hierarchy (Fig. 3–1) provides a separation of functions, which is useful in designing a complex system. Also, it permits evolutionary changes of the protocol since each level acts as a separate module.

There are four levels of protocol standards adopted, and seven defined between data terminal equipment (DTE) and data circuit equipment (DCE). At Level I, there is a physical interface between the DTE and data circuit equipment (DCE). In the past, this interface has been implemented using RS-232. In the future, new versions of the electrical interface such as RS-422 will be employed. At Level 2, there is the link control procedure, and here the prime example is the standard high-level data link control (HDLC). At Level 3, there is the network control protocol. Within X.25, the Level 3 protocol defines the packet formats and the control procedures for exchanging information between a DTE and the public packet-switched data network. At Level 4 there are the user-to-user protocols connecting one DTE to another DTE on an end-to-end level. Such protocols have not yet been standardized internationally, but will eventually focus on terminal-to-terminal communications.

As information moves through the network from one protocol layer to the next, the lower-level protocol accepts all the data and control information of the higher-level protocol and then performs a number of functions upon it. In most cases, the lower-level pro-

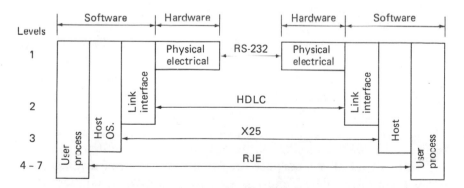

Fig. 3–1. Interface standards have been developed for levels of protocols.

tocol takes all the data and control information, treats it uniformly as data, and adds on its own envelope of control information.

For instance, the host-to-host protocol will take the application data from the user program and add on a protocol header. As this message is sent through a computer network, the subscriber-to-network protocol will add its own control information. Then the link control procedure will add on more information and, in addition, the network will add on subnetwork control fields. It is in this format of messages flowing through a communications network that the concept of the protocol hierarchy is most evident.

3.3 PHYSICAL ELECTRICAL INTERFACE

Early in the history of computers, the EIA organization saw that unless there was a common interface built into all equipment, computer manufacturers would soon be building their own. The Electronic Industries Association (EIA), established a set of Level 1 physical standards which has now come to be known as the RS–232–C interface (international equivalent is CCITT V.24). Basically RS–232–C ensures three things:

1. That voltage and signal levels will be compatible.
2. That the interface connectors may be plugged together (mated) with identical pin wiring and corresponding pin connection.
3. That certain control information supplied by one device must be understood by the other device.

The RS–232–C interface includes 25 pins. Few systems use all of them; however, Table 3–1 gives the most commonly used pins and their functions. It is important to make the distinction between data sent from the terminal (DTE) and data from the data communication equipment (DCE). DTE refers to the terminal: The "terminal" may be either a data entry device, a printer, or a large computer. DCE refers to the data communication equipment (the modem).

The fact that all systems use the same pins does not mean that all devices with RS–232–C interfaces will operate when connected. A connection must have, not only the right pins, but the right infor-

Table 3-1. Commonly Used RS-232-C Signals

SIGNAL NAME	MNEMONIC	PIN	USE
Protective Ground		1	A connection to the terminal's metal chassis.
Transmitted Data	TDATA	2	Outgoing data path from the terminal's point of view.
Received Data	RDATA	3	Incoming data path from the terminal's point of view.
Request to Send	RTS	4	Activated by the terminal to tell the modem to prepare to receive and retransmit data from the terminal.
Clear to Send	CTS	5	Activated by the modem to tell the terminal that it is ready to receive and retransmit data from the terminal.
Data Set Ready	DSR	6	Activated by the modem to tell the terminal that the modem is operational.
Signal Ground		7	Return path for all other signals on the bus.
Received Line Signal Detector	CD	8	Activated by the modem to tell the terminal that the modem has made contact with the modem of the far end, and can sense the carrier.
Data Terminal Ready	DTR	20	Activated by the terminal to tell the modem that the terminal is operational.

mation on those pins. The timing of the clocks (for both DTE and DCE devices), the required control characters, and the proper idle characters are not ensured by RS-232-C.

3.3.1 Handshaking

Handshaking is a process whereby the modem and the DTE device prepare a communications channel for data transfer. Figures 3-2 through 3-4 illustrate the activities on certain lines which must take place before a terminal can transmit and then receive data. This example assumes a normal half-duplex communications channel. The

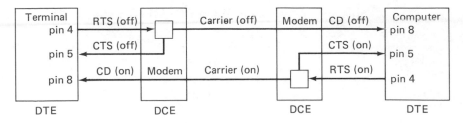

Fig. 3–2. Communications channel in idle state.

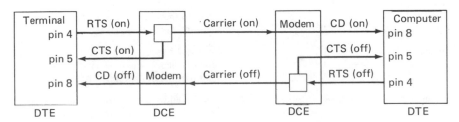

Fig. 3–3. Changing terminal to transmit state.

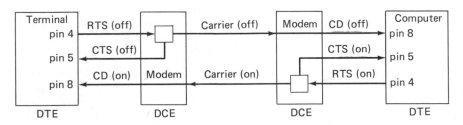

Fig. 3–4. Placing computer in response state.

pin numbers and names are those of the pins on the RS–232–C interface.

Note in Fig. 3–2 the carrier (a continuous signal capable of being modulated to carry information) from the computer to the terminal is on. The modem at the terminal end detects this on pin 8 (carrier detect). Because the terminal wishes to send nothing, its request to send (RTS) pin 4 is off. Therefore pin 5, the terminal's clear to send (CTS), which is driven by the modem, is also off.

Because the computer is seeing no carrier from the terminal, its carrier detect (pin 8) is off. Because it is sending a string of idle characters, its RTS is on and, so is its clear to send (CTS).

In half-duplex systems as presented in Fig. 3-3, the terminal must wait for the computer to indicate that it is ready to accept data before the terminal can send any. When the computer is ready to listen, it turns off its RTS and the computer's carrier signal goes off. The modem at the terminal end sees that the carrier is off and turns the CD off. If the terminal has something to send, it turns on its RTS, which tells the modem to start sending the carrier signal. After a timed wait, the terminal modem assumes the computer's modem sees its signal and tells the terminal that it is permitted to transmit by turning on the terminal's CTS.

In Fig. 3-4, when the terminal has nothing more to transmit, it turns off its RTS. The terminal's modem in turn dops its carrier signal. The modem on the computer side senses this and turns its carrier detect (CD) off. When the computer has something to say, it turns on its RTS, which tells its modem to start transmitting the carrier signal. After an appropriate wait, the computer's modem assumes the terminal's modem has seen the carrier signal and gives the computer the clear to send (CTS). The computer then starts transmitting data.

3.4 LINK CONTROL STRUCTURE

While there are many kinds of link control procedures in use, they share a number of functions in common, including data transparency, connection and disconnection, failure recovery, error control, and sequencing.

3.4.1 Data Transparency

One of the key functions of any link control procedure is to provide for data transparency, which involves either distinguishing data bits from control bits, or providing a means for telling where a particular data block begins and ends without restricting the content of the data blocks. Two general methods for achieving data transparency are in widespread use. The first, known as byte stuffing, is used in the binary synchronous communications (BSC or bisync) protocol. The second, and related method, is known as bit stuffing and is used in HDLC (see Section 3.6).

These protocols work by assigning control meaning to certain data patterns. That is, the beginning of a block is flagged with a certain pattern of ones and zeros. In the byte stuffing case, this is a pair of characters or bits in a message, the sender adds special bytes or bits which the receiver removes, preserving the original data pattern.

3.4.2 Connection and Disconnection

The link control procedure must provide a means for establishing and terminating conversations. Usually this is done through a virtual connection between sender and receiver. At the time the connection is established, the sender and receiver exchange various kinds of identification, which may include the various parameters of the conversation. It must also provide a means whereby the conversation can be disconnected and, more important, reconnected in the event of error. All these techniques are needed so that the link control procedure can operate in case of errors or failure in the link itself or the data equipment at either end.

3.4.3 Failure Recovery

To permit recovery from complete failures of the link or data equipment, the procedure must allow for resynchronization of various communications parameters. This means that the data equipment operating the link control procedure must continuously perform a test to see that the link is operating normally. When the link is determined to be inoperative, both ends must take appropriate measures so they both understand that the link is inoperative, even though it may still be functioning in one direction. Once both ends know the link is down, they can begin to monitor the circuit and, after it is working correctly again, establish a new connection.

3.4.4 Error Control

There are many possible methods for error control over a transmission link, but the most practical approach is to provide an error detection mechanism for each transmitted message and to allow for retransmission by the source. This retransmission can be triggered either by a negative acknowledgment from the destination when it does not receive the intended message, or by a scheme of positive acknowledgments from the destination when it does receive messages. Another alternative is a timer mechanism in the source so that it retransmits messages when it fails to receive positive acknowledgments.

Error detection can be performed by a variety of means, such as simple parity checking, or the more complicated system of cyclic redundancy. When an error is detected, the destination can choose to ignore the block which can then be retransmitted by the source. More details on this subject are discussed in Section 3-12.

3.4.5 Sequencing

While sequencing of traffic is not an essential property of a link control procedure, it is often employed to ensure that traffic leaving a particular link is virtually the same as traffic entering. In a large packet-switching network with many links, the practice is normally not to sequence traffic on individual links since it can be sequenced at the ultimate destination. In centralized networks the opposite approach is often taken. The usual solution to keep data blocks in order is to assign a sequence number to each block for use in detecting missing and duplicate blocks.

3.5 BISYNC PROTOCOL

While bit-oriented protocols represent the wave of the future, the most common link control procedure remains IBM's binary synchronous communications protocol, or bisync, which has been used since 1968 to control transmission between IBM computers and terminals. Since bisync uses special characters to delineate the various fields of a message, it is known as a "character-oriented protocol."

In point-to-point bisync, data blocks have a simple structure. They begin with a special two-character sequence (DLE-STX) and end with a special sequence (DLE-ETX). As the data is being transmitted, a cyclic redundancy checksum is computed and sent at the end of the block (see Fig. 3–5A). The receiver also constructs the checksum, which it compares with the transmitted version to determine if transmission errors occurred. If the two checksums match, the receiver sends a positive acknowledgment and the transmitter can send its next block. If there is a discrepancy, the receiver sends a negative acknowledgment, and the previous block is retransmitted.

For multipoint configurations, the protocol functions by designating one device as master and the other devices as slaves. The master then sends a special block to each slave in turn to determine if the station has any data to transfer. This block also contains an identification sequence, or address of the slave being polled. If the slave is ready to transmit, it sends a positive acknowledgment followed by a data block. If it has nothing to send, it transmits a negative acknowledgment and the polling continues with the next slave in the sequence.

Despite its popularity, bisync operates in half-duplex mode and requires each block to be acknowledged before the next one can be transmitted. These elements reduce throughput as compared to the bit-oriented protocols, which operate in full-duplex mode and permit the transmission of several consecutive frames without requiring an intervening acknowledgment. Also, bisync is cramped by a character-based control format that makes it sensitive to code structure. Since the new protocols, such as HDLC, use bit sequences to convey control commands and information, they provide transparency to code structure.

DLE	STX	DATA	DLE	ETX	CRC

Fig. 3–5A. Character oriented protocols.

3.6 HDLC PROTOCOL

Internationally, the adopted level 2 standard is HDLC, which is similar to IBM's synchronous data link control procedure (SDLC), and ANSI's advanced data communication control procedure (ADCCP). HDLC is actually a family of link control procedures defined by a number of modes and options. For instance, HDLC has two basic modes of operation—normal response mode, designed for centralized systems in which a primary station polls the secondary station; and the asynchronous response mode designed for situations in which either station can transmit at any time.

Also, HDLC is continuously undergoing revision, and there are now other modes such as a balanced mode of operation in which neither station is designated as primary or secondary. The HDLC frame format allows for data transparency by means of bit stuffing. Eight-bit address sequences are also used to control the flow of information (see Fig. 3–5B). For its data link control protocol, the X.25 standard uses a subset of HDLC known as the link access procedure (LAP).

3.7 SDLC OVERVIEW

SDLC is one of the most popular bit oriented protocols. It can be viewed as an envelope in which data are transferred from one station to another across a data communication link. The link can be either multipoint or point to point and involves at least two participating stations. SDLC data links assign two roles to these stations: primary (commanding) and secondary (responding). The primary station is

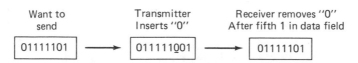

Fig. 3–5B. To achieve data transparency, protocols employ bit stuffing. In bit-oriented protocols (BOP's) such as HDLC, the transmitter "stuffs" an extra "0" after each sequence of five 1 Bits appearing in the data; the receiver deletes any such "0" bit to avoid confusion.

Table 3-2. Common Protocol Characteristics

FEATURE	BISYNC	DDCMP	SDLC	HDLC
Full-duplex	No	Yes	Yes	Yes
Half-duplex	Yes	Yes	Yes	Yes
Message format	Variable	Fixed	Fixed	Fixed
Link control	Control character	Header (fixed)	Control field	Control field
Station addressing	Header	Header	Address field	Address field
Error detection	VRC/LRC	CRC	CRC-CCITT	CRC-CCITT
Control characters	Numerous	SOH, DLE,	Idle flag	Idle flag
Character codes	ASCII, EBCDIC	ASCII	Any	Any

responsible for the data link and issues commands to which all other participating stations respond. The basic unit of information on an SDLC link is a "frame."

Each frame is enclosed in "flags," which are used for frame synchronization and have the binary configuration 01111110. The opening flag serves as a reference for the position of the address and control fields. The closing flag terminates the check for transmission errors. Table 3-2 presents a comparison of protocol characteristics.

3.8 SYSTEM NETWORK ARCHITECTURE (SNA)

Advances in computing, processing and communications technologies have prompted increased interconnection of terminals and communications facilities into networks for distributed access to processing and data-base resources.

A variety of networking applications have been developed for airline reservations, banking, store checkout, process control, and office systems. The network environment requires a master interconnection strategy so that diverse products and applications can share computational and communications facilities while interacting compatibly.

Since its introduction in 1974, IBM's systems network architecture (SNA) has provided the blueprint by which the capabilities of IBM networking products have evolved in an orderly fashion. SNA provides rules for all levels of interaction, from physical/electrical interconnection of computing devices and terminals to meaningful application-oriented processing. The lower layers control only the basic transfer of bits, while the higher layers support meaningful exchange of messages and allow data-base sharing. SNA has also incorporated protocols adopted by national and international standards organizations. This means SNA is compatible with standards such as X.25 public packet switching, high-level data link control and the data encryption standard.

3.9 POLLING EXPLAINED

"Polling" involves the addressing of each terminal on the line by the host computer. In a typical application, the computer polls the first terminal which responds "NAK" if it has nothing to transmit, or "ACK," followed by its message if it has a message to transmit. The computer then polls the next terminal in sequence. If any terminal does not respond, the computer will "time out" and proceed to poll the next terminal. Polling takes place constantly, in round-robin fashion. Outbound messages from the computer are also transmitted to each terminal when it is due to be polled. Typically, the ACK-NAK protocol is also used to verify correct receipt of messages or request retransmission. Terminal polling is only possible if the terminal is smart enough to have an address and able to respond when it reads its address in a message received on the line.

3.10 TRANSMISSION CODES

There are probably as many transmission codes in existence today as there are types of terminals. This proliferation exists partly because computer manufacturers want to ensure the use of their own terminals, partly because of special systems requirements and partly because no standard existed 10 years ago when data communications began rapid development. Transmission codes differ from computer codes only in that certain code values are reserved for terminal control and data formatting. Some transmission codes such as BCD resemble character-oriented computer coding and employ a parity bit for error detection, as is usually done in a computer code. On the

	BITS			$b_7 \rightarrow$	0	0	0	0	1	1	1	1
		$b_6 \rightarrow$			0	0	1	1	0	0	1	1
			$b_5 \rightarrow$		0	1	0	1	0	1	0	1
b_4	b_3	b_2	b_1	HEX	0	1	2	3	4	5	6	7
0	0	0	0	0	NUL	DLE	SP	0	@	P	`	p
0	0	0	1	1	SOH	DC1	!	1	A	Q	a	q
0	0	1	0	2	STX	DC2	"	2	B	R	b	r
0	0	1	1	3	ETX	DC3	#	3	C	S	c	s
0	1	0	0	4	EOT	DC4	$	4	D	T	d	t
0	1	0	1	5	ENQ	NAK	%	5	E	U	e	u
0	1	1	0	6	ACK	SYN	&	6	F	V	f	v
0	1	1	1	7	BEL	ETB	'	7	G	W	g	w
1	0	0	0	8	BS	CAN	(	8	H	X	h	x
1	0	0	1	9	HT	EM	)	9	I	Y	i	y
1	0	1	0	A	LF	SUB	*	:	J	Z	j	z
1	0	1	1	B	VT	ESC	+	;	K	[	k	{
1	1	0	0	C	FF	FS	,	<	L	\	l	/
1	1	0	1	D	CR	GS	-	=	M	]	m	}
1	1	1	0	E	SO	RS	.	>	N	~	n	~
1	1	1	1	F	SI	US	/	?	O	_	o	DEL

Fig. 3-6. American standard code for information interchange (ASCII).

other hand, Baudot, for example which is a five-level code, has no provision for parity or the detection of errors. A chart of the code values of ASCII (Fig. 3-6) has been included since its the most widely used. ASCII represents an effort to establish a uniform coding scheme for all terminals: in practice, however, it has not been implemented on the scale the designers envisioned for it.

Of the other existing codes, the more widely used ones are the extended binary coded decimal interchange code (EBCDIC), the five-bit Baudot code found in old teleprinter equipment, the four of eight code, the IBM punched card Hollerith code, the binary coded decimal (BCD) code, and the six-bit transcode.

EBCDIC is an eight-level code similar to ASCII, except that while ASCII uses its eight level for parity bits, EBCDIC uses it for information bits, thereby extending the range of characters to 256.

3.11 CODE AND SPEED CONVERTERS

A combination of speed-and-code converters permits the user to satisfy two or more dissimilar communications requirements with a single terminal, thereby reducing both the number of terminals necessary and capital investment (see Fig. 3–7). Consider an organization with many ASCII–code eight-level Teletype models 33 and 35 used on the TWX network and in timesharing and with some older, five-level Baudot Teletype models 28 and 32 used on Telex and international-record-carrier networks. It is desirable to standardize on the models 33 and 35 because they are more widely used and more readily available. A combination of speed and code converters will make them usable on Telex-and TWX-like networks as well as in timesharing. The ASCII even-parity serial code of 11 bits (100 words per minute) is converted to a Baudot serial code of 7.5 bits (66 words per minute) and vice versa. Since most ASCII terminals can be interfaced by a converter, the more versatile CRT terminals can send messages through Telex or TWX-like networks.

3.12 TREATMENT OF ERRORS

Many data transmission codes use an extra bit, called a parity bit, in each character for checking purposes. This is added so that the total number of 1 bits in the character transmitted will be an odd (or in some codes, even) number. The receiving machine detects whether there is an odd number of 1's. If not, it knows that noise or distortion on the line has lost or added a bit. It either notes this error or calls for a retransmission of the data. This is sometimes

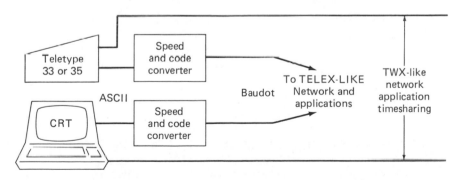

Fig. 3–7. Combining speed and code conversion.

called vertical redundancy checking. Unfortunately, noise on the communication line sometimes changes more than one bit, and this lessens the effectiveness of parity checking. If two 0 bits are both changed to 1's, the parity check will not detect this.

Vertical parity, in which one extra bit is added to each character, will detect or correct about 85 percent of all errors transmitted, and will result in an average of one undetected error for every 700,000 bits transmitted. The efficiency of circuit usage, however, is somewhat degraded, since one extra bit per character must also be carried over the line with the data. In a five level code, this would mean a reduction of about 20 percent in throughput; in higher level codes, of course the efficiency would not be reduced as much.

Double parity checks (Fig. 3–8) involve both checking the vertical

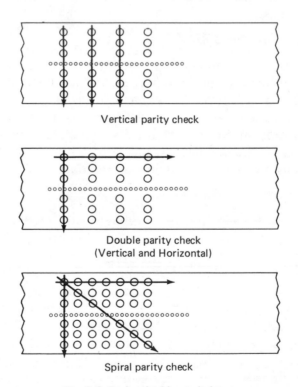

Vertical parity check

Double parity check
(Vertical and Horizontal)

Spiral parity check

Fig. 3–8. Parity checking techniques.

parity as well as sending an additional check character per block of information. With this scheme, error control becomes much more effective, so that only one undetected error will occur, on the average, for each 10,000,000 bits transmitted.

Sophisticated error control schemes are in use for especially critical applications; however, their description is highly involved and so will not be fully covered here. Perhaps the most common of these is spiral parity, in which a parity check, over and above the double parity scheme, checks diagonal rows of bits for odd or even parity. Such a system assumes that only one error in every 100 million bits transmitted will remain undetected.

3.12.1 Redundancy Checks

Many error schemes include a block check character at the end of the transmission. This character is used for error checking for the whole block of data that was sent. This type of error detection is called a cyclical redundancy check (CRC). In CRC error detection, the transmitting device performs a mathematical computation on the binary value of the bits in the blocks to be transmitted, and adds the results to the blocks of data. At the receiving end, the same computation is done. If the result is the same as the one sent, then everything is fine. If not, the whole block of data must be retransmitted.

3.12.2 Error-Correcting Codes

Automatic correction can take a number of forms. First, sufficient redundancy can be built into the transmission code so that the code itself permits automatic error correction as well as detection. This is referred to as forward-error correction. To do this effectively requires a large proportion of redundancy bits. Codes that give safe forward-error correction are, therefore, inefficient in their use of communication line capacity. If the communication line permitted the transfer of information in one direction only, then they would be extremely valuable. However, most systems use half-duplex or full-duplex links. Forward-error correction becomes advantageous when the number of errors is so high that the retransmission of data necessary would substantially degrade the throughput. On higher-speed links, the argument for forward-error correction becomes stronger because the time taken for reversing the direction of transmission is equivalent to many bits of transmission. This time is relatively high on wideband links and on half-duplex, voice-grade links with high-speed modems.

3.12.3 Loop Check

One method of detecting errors does not use a code at all. Instead, all the bits received are retransmitted back to their sender, and the sending machine checks that they are still intact. If not, then the item in error is retransmitted. Sometimes referred to as a loop check or echo check, this scheme is normally used on a full-duplex line or on a continuous loop line. A loop check is most commonly found on short lines and in-plant lines where the wastage of channel capacity is less costly. It gives degree of protection that is more certain than most other methods.

3.12.4 Effect of Errors

On many data transmission systems, the control of errors is of vital importance. On some, it is not of great significance. Some systems transmit vital information, like accounting data, financial figures, military orders, encoded medical data or programs. These must be word perfect (or bit perfect). Other systems transmit the information that an operator is keying in at a terminal. The operator is likely to make far more errors than the data transmission line, especially if he is an unskilled one-finger operator. In this case, there is no point in worrying too much about the line. Accuracy controls can be devised for human input. On many systems a tight network of controls is necessary to stop abuse or embezzlement. It is also important to ensure that nothing is lost or double-entered when hardware failures occur on the system or when switchover takes place.

In designing a system, it is important to know what error rate is expected. Calculations should then be done to estimate the effect of this error rate on the system as a whole. On some systems the effect of infrequently occurring error is cumulative, and it is in situations such as these that special care is needed in eliminating errors. For example, if messages cause the updating of files, and an error in the message causes an error to be recorded on the file, then it is possible that as the months pass the file will accumulate a greater and greater number of inaccuracies.

3.13 FRONT-END PROCESSORS

The most important use of programmable communications processors today is front-end processing, in which the processor handles all message control activities, and performs enough preprocessing to relieve the call of the communications housekeeping activities.

The concept of front-end processing essentially involves off-loading or removing the data communications control function from the central processing unit and setting it up as an external, largely self-contained system. This decentralized approach permits both the communications and central processors to perform their primary functions in parallel and with little interference. Data is passed between the processors only when necessary and with a high degree of efficiency.

A typical front-end processor might control a hundred or more communications lines of varying speeds and types attached to a large number of diverse remote terminals. The front-end processor would ideally assume all terminal, line, buffering, and message control functions, permitting the central processing unit and the user application programs to treat the communications network as just another high-speed, on-line peripheral device. Because the front-end processor is essentially a programmable computer, it can be programmed to perform an almost limitless variety of functions. But in its role as external controller of a centralized data communications network, the specific functions generally programmed are those that relate to data and message control.

3.13.1 Selection and Evaluation

Front-end processors deserve careful investigation because of their many apparent advantages over hard-wired communications controllers. Such investigations should include as many probing questions as possible, because there are potentially serious pitfalls to be avoided.

One potential problem is the question of overloading the front-end processor, with the resultant loss of data. Sophisticated data and message control programs will consume large quantities of the front-end processor's computing and memory facilities, just as they do in a centrally-based communications system. Since many front-end processors are based on minicomputers, the possibility of overloading is all the more real. A tendency toward overloading can easily negate any apparent advantages of expandability and growth potential.

Another serious question is that of software. The body of software required for terminal control, line control, and message control activities, not to mention application-oriented pre-processing, is unquestionably complex. It is also vital to the operation of these systems. The prospective user must determine whether or not the supplier is capable of supplying this software, at what level of completeness, with what assurance of bugfree stability, with what chances of interfacing smoothly with the central system software, and with how much installation assistance.

Another buyer's tip is to look for the word "turnkey." Turnkey installation of front-end processors usually means that the supplier takes on full responsibility for hardware, software, and interfaces required to essentially "plug in" his product. From a user's point of view, this approach is highly desirable, since it can save him money, time, and aggravation.

When considering a front-end processor from a source other than the supplier of the central computer equipment, the buyer should insist on receiving concrete performance data, drawn from installed systems, to substantiate the supplier's claims. The buyer should beware "if" the supplier refuses to back up his claims with actual case studies. As further evidence of proven performance, the buyer should personally contact as many previous users as possible, probing not only for their degree of satisfaction, but also for the extent to which the installed systems reflect his own intended system design and objectives.

4
TERMINAL TECHNOLOGY

4.1 CRT DISPLAY TERMINALS

The most significant consideration affecting terminal selection is matching the functional capability of the terminal to the functional requirements of the application. The best evidence of this importance lies in the variety of special-purpose terminals developed for such applications as airline reservations, on-line banking, and brokerage transaction processing. The application primarily sets the requirements of speed, permissible error rates, input/output medium, information codes, and format storage in a terminal.

CRT display terminals employ a TV-like display screen to record and display data. They have a keyboard similar to that of a typewriter (see Fig. 4-1). When a key is depressed, the associated numeric or alphabetic character appears on the screen. The character code may be simultaneously transmitted or it may be stored for subsequent transmission as part of a word or a complete message. Typically, 1000 characters may be simultaneously displayed on the screen, and character transmission rates are in the range of 110-to-1200 characters per second.

CRT display terminals are widely used in inquiry-response applications as "electronic blackboards" to provide rapid access to data stored in computer systems. These terminals are preferred due to their ability to selectively alter the displayed data.

To provide data processing services, an intelligent terminal should provide operating system, arithmetic and logic functions, as well as handle local files. It should also be able to run application programs

Fig. 4-1. CRT terminal from Hewlett-Packard. (*Courtesy Hewlett-Packard.*)

written in a high-level language, and to communicate with the host processor over a telephone line.

To fulfill these requirements, most intelligent terminals employ hardware modules consisting of a control unit, an alphanumberic keyboard, and a display screen.

4.1.1 Control Unit

The control unit (Fig. 4-2) directs and services the needs of the terminal. The control unit usually is integrated into the terminal, but in some cases it's a separate unit. Generally, separation is used when the terminals are configured in a cluster, where two or more termi-

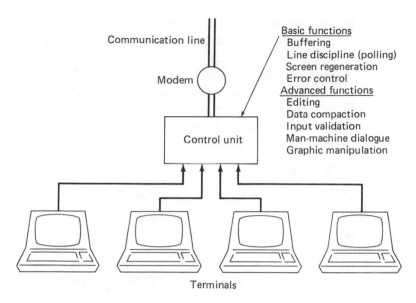

Fig. 4–2. Intelligent terminal control unit functions.

nals are handled by one controller. The control unit usually contains a microcomputer supported by a random access memory (RAM). The RAM is used to handle application programs, data files, system utilities and, in some cases, the operating system.

4.1.2 Keyboard Structure

The keyboard is generally laid out in typewriter-style, although some terminals with advanced text processing capabilities may furnish a slightly modified version. Associated with the keyboard can be a separate keypad which allows entry of numeric data. All keyboards provide separate keys for positioning the screen cursor, editing local data, and keys which can be programmed for special functions. The screen cursor points to the position on the screen where the next character will be displayed. Cursor movement can be handled manually from the keyboard or can be under program control. Normal cursor movement is up, down, left, right and home.

Some terminals also provide protected fields and scrolling. Protected fields are used to retain certain data fields which are used repetitively. For example, a protected field would be a title such as name, address, etc., after which the operator enters information. When the transmit key is depressed, only the variable data is transmitted. Most terminals provide a scrolling function. Scrolling allows the operator to roll page information retained in a scrolling buffer back onto the screen where it can be checked or modified.

4.1.3 Display Schemes

Most terminal displays are cathode-ray tubes (CRTs) which employ a dot matrix to form the character. The screens are rectangular with their measurements given in diagonal inches. Screen sizes vary from a few inches to about 15 inches. The screen's capacity is the maximum number of characters that can be displayed at one time. Most manufacturers have standardized on 1920 characters which is derived from a format of 24 lines by 80 characters per line. An increasing number of vendors are also offering 132-character-per-line displays. This corresponds to a typical computer printout, and can be used to check format. Of course, the more characters displayed per line, the harder it is to read.

4.1.4 Terminal Considerations

There are two conflicting interests and tradeoffs in terminal selection; namely expense verses line utility. Most low-cost terminals are start-stop, with no buffers. Buffered, synchronous terminals are more expensive but give better line utilization. Where the lines are short and inexpensive (e.g., within one city), efficient line utilization is of little importance. When a dial-up line is used, normally there is only one terminal on the line and start-stop operation will often be good enough. A somewhat higher character transmission rate could be obtained with synchronous transmission. The other main advantage of synchronous transmission is that the error rate can be less. Extremely good error control can be achieved with high-order error-detecting codes and a buffer in the sending machine so that retransmission can be automatically requested. The expense of synchronous operation is diminishing as the cost of logic circuitry drops. At the same time, the reliability of logic circuitry is substantially increasing. All the logic for a buffered, synchronous terminal can now be constructed on one large-scale integration chip, which are low cost when mass-produced. The use of buffers is the only way to guarantee a good response time on a multidrop line, and to make transmission from the buffer as fast as possible, it would be worthwhile to use synchronous transmission.

A terminal on a polled line (Fig. 4–3) must recognize its own address and respond to the polling signals. The logic for this process is contained in the terminal control unit, buffer, plus the ability to code and decode error-detection characters in messages and to retransmit or request retransmission of errors.

The terminal may use full-duplex or half-duplex transmission. It

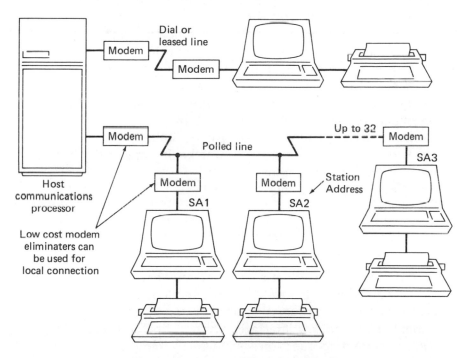

Fig. 4–3. Typical terminal configurations.

may be designed so that data can travel in both directions at once, although it is more common for data to travel in one direction while only control signals travel simultaneously in the other. Many terminals are not capable of transmitting and receiving at the same time—in which case, there is no point in paying extra for a full-duplex line for them. In some countries a full-duplex line costs the same as a half-duplex line, and so a terminal that has only half-duplex capability wastes some of the available bandwidth. The following list can serve as a guideline for CRT terminal selection.

- Can it display enough characters?
- Are the displayed characters large enough?
- Are the characters easy to read?
- Is it flicker-free?
- Is the image bright enough?
- Is the image suitably protected from external glare?
- Is the display rate fast enough?
- Can it handle vectors or other graphic features?
- Is a modem or acoustical coupler built in?
- Is full-duplex or half-duplex transmission used?
- What cursor movements are possible?
- Does it have selectable tabs or other formatting features?
- What editing capabilities are available?
- Can different codes be used, for flexibility?
- Is there a nonprinting feature for keying-in security codes?
- Is there a cryptography feature?
- Is there a physical lock and key?

4.2 PRINTING TECHNOLOGIES

In spite of the promise of a "paperless society," printers and other hard copy devices have continued to advance both in the market place and in technological innovations. In a modern communications system, either for data processing or office applications, printers have become one of the key interfaces between operator and machine.

For a printer to function satisfactorily requires more than the making of marks on paper by a printing mechanism; it needs a paper transport mechanism, electronic controls, and a means for inputting information to the printer, such as a keyboard or some sophisticated communication devices.

Printers can be categorized into impact and non-impact printers. The impact printers use the kinetic energy of a rapidly moving hammer, character ball, or other actuators to drive an ink ribbon against paper for ink transfer to the paper (so-called front printing). The printing can also be done by impacting the paper against the ribbon backed by a platen (so-called back printing). Since the impact force of the printing element can be transmitted through several sheets of paper and carbon paper, it can simultaneously generate multi-copies—a key advantage of impact printers.

"Daisy wheel" printers were the outcome of a major innovation in serial impact printer mechanisms. As shown in Fig. 4–4, the daisy

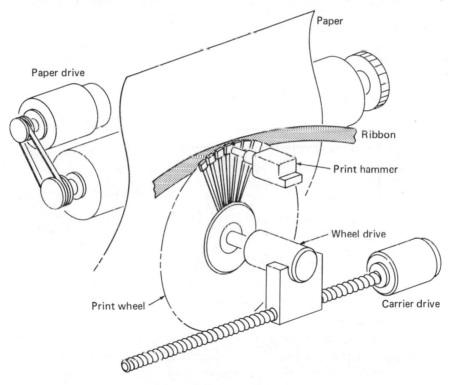

Fig. 4–4. Daisy wheel printer.

wheel printer uses fully formed characters on the tips of flexible bars arranged in a daisy-wheel fashion. This print wheel rotates and stops in front of an electromagnetic actuator to select a character to print. The daisy wheel with its light weight and simple rotation enables the printer to operate at a burst speed as high as 70 cps.

Due to the need of increasing the flexibility of electronic font and form generation, wire matrix printers have made rapid strides. Wire matrix print heads (as shown in Fig. 4–5) have individually activated print wires usually arranged in a vertical array or staggered arrays. As the print head moves across the page, the print wires form characters in matrix boxes. With proper electrical controls, the wire matrix printer is capable of all-points-addressable printing to print graphs or images in addition to character printing. It is also possible

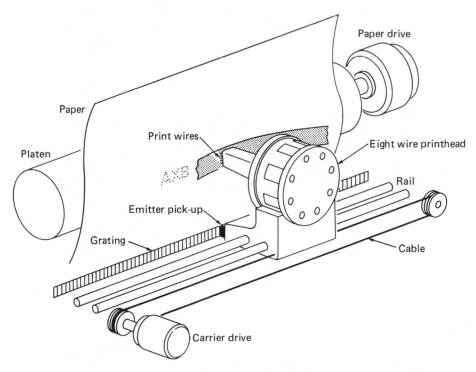

Fig. 4–5. Wire matrix printer.

to obtain multi-copies. The multi-copy capability and reliability have made matrix impact printers attractive.

Even though the print-speed distinction between a serial and a line printer has become less distinct and overlapped, a line printer is historically considered as printing a line of characters at once. Figure 4-6 shows a band line printer with an array of fixed hammer banks. The printing is accomplished by a set of fully formed characters arranged in a chain or band moving in front of the hammer at a high speed to be addressed by the hammers.

The band printer is capable of high-speed printing. The speed limitation for good quality printing is a function of the hammer flight times. Any variation in the hammer times results in misregistration and slur.

Matrix line printing can also be performed by a drum configuration. A "bar helix" line printer uses rotating helical wires wrapped in a drum and a horizontal hammer bank with sharp hammer faces to form print dots. By properly timing the impacts of the line ham-

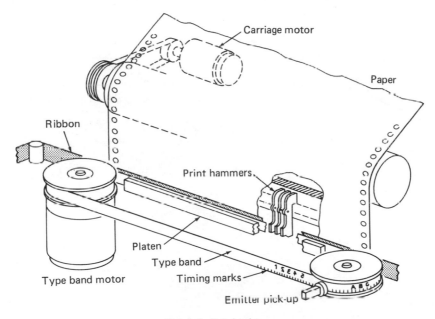

Fig. 4-6. Band printer.

mers, the intersection of the helical wires and hammer surfaces creates a line of dots as the paper moves vertically. Because of the continuous paper motion, bar helix printers have the advantage of high speed. However, bar helix printers usually do not have superior print quality.

The electrophotographic printers (laser printers) (Fig. 4–7) are based on the same basic principle as a copier machine. This non-impact machine uses an optical system to image a document onto a rotating photoconductor drum. The charge image on the drum is toned by charged toner and transferred to plain paper electrostatistically. The transferred toner on the paper is fused by a hot roll or radiation for the final output. A laser printer uses a modulated laser beam which scans across the photoconductor drum to paint the charge image on the drum point by point.

An ink jet printer can be operated either in a continuous synchronous form or in a drop-on-demand mode. As shown in Fig. 4–8, a continuous ink jet forms a stream of ink droplets about 2 to 4 mils

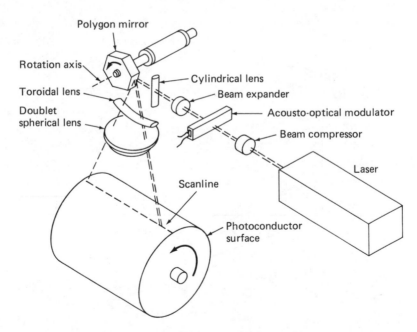

Fig. 4–7. Laser printer.

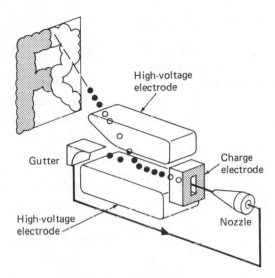

Fig. 4–8. Ink jet printer.

In diameter by pumping ink through a nozzle. These ink droplets are charged to different voltages by a voltage charge electrode and then deflected to different heights vertically onto the paper. Unused droplets are collected in a gutter. The ink jet nozzle head moves horizontally to write a line.

The application demands and technology innovations have continually advanced the price performance of both impact and non-impact printers. One clear trend is the growing popularity of all-points-addressable (APA) printers. The advantages of APA for forms generation, fonts, images, and graphics have been the key selling points of non-impact printers. Impact matrix printers have also moved impact printing into the APA domain. The availability of low-cost microprocessors has made the current printers more intelligent in terms of machine controls, self-diagnosis, and communication with computers and other terminals.

The cost of operating a printer subsystem includes the basic machine, the maintenance cost, and the cost of consumables (paper and ribbon). A sophisticated user has to understand the total printing cost before choosing a particular printer. He also had to appraise the tradeoffs in a printer subsystem for his particular application.

4.3 GRAPHIC TERMINALS

In this technology, the choices for consideration are stroke-writing, storage tube and raster graphics.

Basic operations available with every stroke-writing type graphic display are "move" and "draw". The "move" command moves the beam from one location to another without producing any light with the beam turned off. "Draw" moves the beam in a straight line from one point to another with the beam on, thus producing a straight line of light on the screen. Curves may be approximated by a series of end-to-end short lines or other types of draw commands may be available on some stroke systems to draw curves, circles, and ellipses directly. The major point to remember is that stroke-writing graphic displays can draw a line directly from any random point on the screen to any other random point. The system produces high resolution drawings because it draws the pictures directly from point to point (Fig. 4–9) on the screen.

The storage-tube or DVST type display is a relatively complex structure, but the cost of the tube is balanced by savings in the support electronics (Fig. 4–10). There is no constraint in the amount of time required to trace an image. The amount of data that must be stored for quick-access refresh is reduced to a minimum, yet extremely complex images can be created.

Raster type graphic displays draw lines as a series of pixels, which

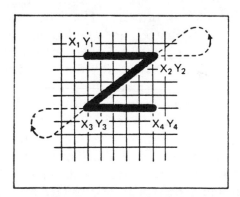

Fig. 4–9. A stroke writing sequence for displaying graphics.

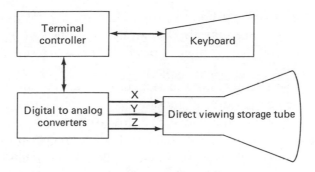

Fig. 4-10. A typical DVST graphics system.

makes diagonal lines appear as stairsteps and consequently limits resolution. Raster terminals are generally less expensive than stroke-writing types and they provide full-color capabilities. The information on the raster screen is typically refreshed from a "bit map" where each bit in the map, which may be several bits deep to provide color, corresponds to a picture element (pixel) on one of the scan lines. See Table 4-1 for a comparative summary of the various types of graph terminal tradeoffs.

Table 4-1. Comparison of Graphic Display Technologies

GRAPHICS DEVICE	LINE QUALITY	SCREEN DYNAMICS	DISPLAY CAPACITY	AREA FILL	COLOR QUALITY
Stroke writer	Excellent	Excellent	Fair	Poor	Fair
Storage tube	Excellent	Poor	Excellent	Poor	Poor
Raster scan	Fair	Fair	Excellent	Excellent	Excellent

4.4 FACSIMILE TECHNOLOGY

In many business applications, facsimile is used to replace long-winded and costly phone calls. In this technology, the thought that goes into communicating occurs before transmission. Messages and sketches are composed before rather than during telephone "conversation." Facsimile may also be employed in place of teletype message transmission or as a higher speed alternative to the mails. Such use can replace typing operations (either for teletype or the mails), where hand written memos and sketches serve as input documents.

One of the key features of facsimile is the ability to operate in an unattended mode. Unattended systems coupled to ordinary telephone networks can send/receive hundreds of pages automatically, using auto-answer phone interfacing. Some units even allow off-hour transmission on an unattended basis via automatic telephone dialing attachments. Positive feedback provides the sender with confirmation that the document has been transmitted and copied at the remote location.

Internationally standardized facsimile machines are divided into three groups in terms of transmission speeds: Group 1 machines, which are capable of transmitting information in six minutes; Group 2 machines, in three minutes; and Group 3 machines, in one minute. The main factor which differentiates the respective group machines is their transmission methods.

Since the facsimile picture signals include signals of frequencies lower than the voice band (300 Hz − 3400 Hz), it is necessary to modulate the facsimile picture signals so they can be transmitted via a telephone line. For G1 machines, frequency modulation (FM) is used. The FM system gives a different frequency to each black and white picture element to be transmitted, as illustrated in Fig. 4–11.

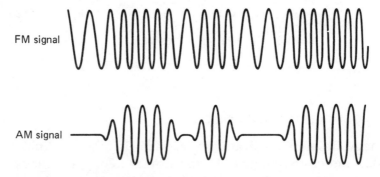

FM signal

AM signal

Fig. 4–11. AM and FM facsimile signals.

The duration of each high or low frequency carrier is proportionate to the length of the black and white picture element. The AM system used in G2 machines defines the white and black picture elements as the largest and smallest amplitudes of a carrier or vice versa.

The standardized G1 FM system is capable of transmitting a volume of picture elements equivalent to 180 scanning lines per minute. It is technologically possible to make the transmission time shorter, but this results in decreased resolution and thus lowers the quality of transmitted information.

However, because G2 machines are required to transmit two times more information (picture elements equivalent to 360 scanning lines per minute) than G1s, they decrease resolution. For example, in cases where black is defined as the smallest amplitude, the black picture signals may not be transmitted. To prevent this, the G2 machine employs phase modulation (PM) technology along with the AM system.

The PM system inverts the phase of the largest amplitude occurring after the smallest amplitude (Fig. 4–12). (a) indicates the phase of an AM signal while (c) indicates this phase inverted to 180 degrees in comparison with that in (a). (b) indicates the AM-PM signal whose different phases appear alternately. By alternately changing the phases, the black signal (usually the smallest amplitude) is always detected. If the black signal is very short, and if the phases do not change, the receiver is likely to recognize the two white signals (the largest amplitudes) as one continuous white signal, while failing to detect the short black signal.

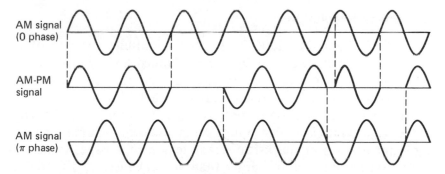

Fig. 4–12. PM system comparison used for G2 fax machines.

The transmission system for G3 machines employs a digitized system as opposed to the analog system used in G1 and G2 machines. Facsimile picture signals are considered redundant if black or white signals are prolonged. Because they are capable of reducing this redundancy by coding techniques, G3 machines are standardized to transmit information in one minute.

A method of reducing redundancy is through the use of the modified READ coding system. This system is based on the fact that black and white elements on one portion of a scanning line are often similar to that on the previous scanning line. This digital coding system codes only those black and white element combinations which differ between two neighboring scanning lines.

4.4.1 Stored Switching System

When connected to a computer system, the facsimile machine can be used as a terminal to retrieve image information stored in the computer's memory. Through the use of a minicomputer system and OMR (optical mark reader), on which several addresses are recorded, a sender can transmit the same information to several receivers one after another. The information to be sent is first stored in a memory and then transmitted in turn to each address. In data processing, facsimile machines are being used as input/output terminals. Facsimile machines are expected to be used increasingly in office automation, especially in image communications.

4.4.2 Image Resolution

Resolution is an important factor in determining the size of the smallest characters that can legibly be scanned and printed, as well as the sharpness or quality of the facsimile output. Both horizontal and vertical resolution parameters are included. Horizontal indicates how many times a scanner or printing element stops along the width of the page. Vertical refers to how many times the scanner or printing element stops or marks down the length of a page. Generally, horizontal parameters are larger than vertical factors, although in many common cases the two may be equal. One scan line consists of one row of picture elements plus one blank row before the next row of picture elements. A resolution of 200 1pi horizontal by 200 1pi vertical will have picture elements of 1/200th of an inch each. It takes a combination of the horizontal and vertical scanning "stops" to complete the construction of a facsimile image.

It should also be noted that resolution is related to transmission time; decreasing the line resolution will decrease the time and cost necessary to transmit the document. Considering today's facsimile techniques (including complex technical factors related to line width, scan sweeps per second, document feed speed, etc.), the use of a system engineered to work with 1/16" high characters will require a higher resolution, and increase the time and communications expense to deliver a document. The time required will be three to four times that associated with 1/8" high characters.

The transmission of tone shades requires the scanning of a large number of points or elements. This entails either the use of more costly high-speed lines, or a longer transmission time. Tone transmission also requires terminal resolution capabilities of 200 lines per inch or higher.

4.5 INTEGRATED WORKSTATIONS

Integrated workstations are keyboard/display devices that permit the creation of integrated text and graphic material on the display screen for eventual printout or communication with another system. These workstations are usually self-contained, and have dedicated files and simplified programming via symbolic keys and menu or symbol selection from the screen.

The display can present characters in several sizes and in different type fonts, and the screen usually may be split into two or more individually controllable work areas. The keyboard is a conventional typewriter keyboard, but there usually are auxiliary keyboards and other devices that can be used to add data or move it on the display screen, such as a light pen, track ball, or "mouse." Externally, the workstation may look like a conventional CRT, but the capabilities are much greater.

5
NETWORK MANAGEMENT

5.1 NETWORK CONTROL DESIGN CONSIDERATIONS

Many existing network control facilities have been constructed in a piecemeal fashion, often for obvious financial reasons, but also due to application growth at new locations and availability of new equipment. While this fragmented situation is less desirable than designing an entire network control system at the onset, it certainly is not hopeless.

Naturally, the kind of network control equipment required will be dictated to a great extent by the number of lines you have and plan to have. If you have only a few lines then it is difficult to justify expensive automatic monitoring, testing and switching devices. On the other hand, if you have a few hundred lines then it will cost you more not to have this kind of equipment.

One underlying consideration which plays over all others is: How much are your data worth? Or, how long can you afford for a line or group of lines to be down? If your network is primarily for record-keeping and subsequent analysis, failures are not so vital and costly since lost or delayed records can be reentered and retransmitted. On the other hand, if your network is an interactive transaction system for bank tellers, airline reservations, travel agencies, or perhaps a time share operation, then downtime might cost thousands of dollars per minute. For these operations a modest investment in capital equipment begins to look very economical.

5.2 ELEMENTARY MONITORING

With only a few lines you could probably get by with VF patching and some degree of monitoring. A VF patch jack is a device which is placed on the telephone line between the modem and the telco line. The "normal-through" connection simply allows the signals to pass through the patch. The double-pointed arrow (Fig. 5–1) represents the normal-through connection. Barrels on the front allow the insertion of plugs for performing tech control functions. Note that the monitor barrel's connection is tied to the normal-through: thus insertion of a plug "picks up" the signal and connects it to a monitoring device while still allowing the signal to flow uninterrupted.

Figure 5–2 shows two such patch jacks connected with a patch cord. On the left, the plug is inserted into the modem barrel of the jack. Note that it breaks the normal-through connection, thereby routing the modem 1 signal onto the patch cord and connecting it to the jack. There, the plug is inserted into the telco barrel thus breaking the normal-through between modem 2 and telco line 2 and routing the modem 1 signal onto telco line 2. Whether patching or switching, analog or digital, the basic principle is the same and it is one of the most fundamental concepts in network control.

You might want to combine your monitoring access with a restoration capability as shown in Fig. 5–3. Here you have a simple application of standard A/B switching or fallback switching which enables you to "spare" the data terminal equipment (DTE). It is even possible to get A/B switching with an adjustable audible alarm on RS–232C leads such as Carrier Detect (CD). Thus, if CD goes

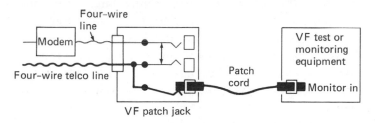

Fig. 5–1. Monitoring through a VF patch jack.

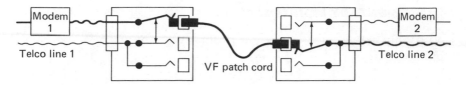

, **Fig. 5-2.** Patch cord connects modem 1 to transmission line 2.

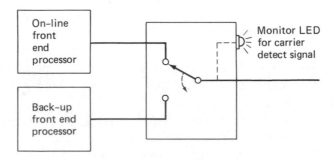

Fig. 5-3. Typical A/B (or fallback) switch.

down (indicating that either the Telco line is lost or the modem has a problem), the operator would be immediately notified.

5.3 REDUNDANCY AND COMPATABILITY

Most telcom planners, including this author, feel it is wise to give preference to equipment which has built-in redundant parts for critical functions, such as power suppplies and control logic, so that if one goes down the other takes over immediately. For major electronic switching systems such as a matrix you should get front panel patching for backup so that if the matrix goes down you can patch across it. It makes little sense to spend a quarter of a million dollars for a swtiching system and then not spend a few thousand more for this kind of redundancy—no matter how reliable you think the equipment is.

There is a large tendency in the business towards standardization. Nonetheless, the compatability of your equipment with equipment you already have of plan to obtain is always an issue. It may be worth your while to call in a consultant who specializes in system design and knows pretty much what fits what. On the other side of this matter, be careful not to make compatibility with existing equipment too much of an issue because in some cases it may be both cheaper and more efficient in the long run to discard your present equipment and configure whole new network control facility.

5.4 NETWORK TESTING

The integration of voice and data communications require test systems that can perform fully integrated test functions. Trying to use test sets designed for older systems to manage modern integrated networks is like trying to perform open-heart surgery with a hunting knife. We need a planned system approach for management and testing functions rather than a fragmented approach.

Modern telecommunications switching systems incorporate centralized diagnostic and test functions within the switch design. With the aid of microprocessors, even nontechnical persons can perform routine maintenance functions from a central office.

Centralized diagnostic facilities make it possible to reconfigure network traffic patterns and to avoid trouble spots without interrupting the flow of information being transmitted. In addition, signal analyzers and datascopes permit testing of individual data lines to isolate malfunctions with pinpoint accuracy.

5.5 CENTRALIZED TROUBLESHOOTING

Central diagnostics troubleshoot and control network equipment from a centrally located test center. This allows the network operator to diagnose and pinpoint failures as well as disable faculty network components and enable backup equipment, all from a central test center miles from the trouble.

The amount and kind of equipment in any test center depends on network characteristics, including architecture, data transmission format, the number of network lines as well as the training and experience of the personnel running the center. The function and scope of any network test center can range from simple line patching facilities that allow the operator to do routine loopback testing and alternative routing, to computer controlled testers that monitor line jitter, phase hits, bit error rate, handshaking, and answerback performance of remote equipment. Using remote answerback facilities, the computer can enable the appropriate test equipment for any situation, monitor the results, and activate backup systems. Operation restoral can often be accomplished so rapidly that the network user is completely unaware that an equipment failure has occurred.

The expense of installing diagnostic restoral hardware must be weighed against the urgency of getting the remote site back online. It is clearly inadvisable to install backup modems and terminals at sites that are used only for sending monthly billings or inventory statements. On the other hand, sites involved with the transfer of large amounts of money are critical; downtime can cost thousands of dollars.

Most networks, however, can safely withstand a few hours of downtime. Since most modems and terminals are not self-repairing, many central diagnostic schemes choose the less expensive alternative of isolating faulty equipment to prevent interference with further network operation. Equipment servicing then takes place at a more convenient time.

More comprehensive schemes—involving diagnostics all the way to the remote terminal—assume a certain uniformity. That is, many vendors offer a complete diagnostic software package that runs on the user's mainframe. This is often contingent on the user buying all network components, including software, from the same vendor; however, one vendor often cannot supply all the necessary equipment for a complex network.

The telecommunications manager is responsible for the operations and maintenance of network switches and trunks. This responsibility includes responding to complaints from network users, administrating network features, managing switch data base modifications as well as monitoring and controlling traffic in real-time, and ensuring that the network is providing the desired grade of service in the most cost-effective fashion. These tasks are further complicated by multiple network nodes and numerous linking trunks, all of which are geographically dispersed.

5.6 NETWORK CONTROL CENTER

The network control center (NCC) provides network performance monitoring and centralized maintenance to ensure maximum availability of network facilities. This is accomplished by providing continuous on-line performance monitoring of switch nodes and providing alerting mechanisms to indicate critical conditions. Another level of insurance can be attained by providing the NCC operator with the ability to detect, diagnose, and localize network problems to a particular switch or trunk so that corrective action may be taken. The NCC can also provide mechanisms for the coordination of maintenance activities and ensuring management control through the use of reports.

5.6.1 Alarms

In a centralized control center environment, the operator must be made aware of critical maintenance conditions rapidly. The network switches remote call maintenance information to the network control center. However, in a multi-switch network, the volume of maintenance information can be so large that critical messages get overlooked. To ensure that critical conditions are not overlooked, an NCC alarm feature can identify messages requiring immediate operator attention.

Every message from each switch in the network can be transmitted to the NCC and compared with pre-defined alarm conditions. If there is a match, an alert is immediately generated. Further, network alarm status information can be dynamically displayed via color graphics, giving the operator an overview, at a glance, of current network alarm status. This avoids the traditional deluge of rows and columns of numbers. A typical graphics display consists of a geographic representation of all nodes and trunks in the network. Immediate recognition of alarm conditions can be assured by changing the node symbol of the affected node from green to flashing red and sounding an audible alert which can only be silenced by operator acknowledgment. The actual message which triggered the alarm can be printed on a specially designed alarm printer.

5.6.2 Automatic Trunk Testing

The NCC provides a centralized capability to automatically or manually test each trunk in the network for noise, as well as gain, and report these results to the operator in an efficient, organized manner. Trunk testing is necessary, because vendors of transmission facilities (common carriers) generally do not test the trunks, so there is no way to assess their performance. Furthermore, because trunk costs usually dominate telecommunications costs, it is desirable to lease the fewest trunks possible for a required grade of service. Thus, if a trunk is defective, not only is an expensive facility going idle, but service is degraded.

5.6.3 Maintenance and Management Reports

The network control center can provide two powerful trouble tracking mechanisms: an alarm log and trouble tickets. These computerized facilities support the network operators and their management by simplifying and assisting in trouble follow-through.

The alarm log is generally a computerized form automatically generated each time an alarm is detected by the NCC. Included in the format are the date and time of occurrence, switch identification, message type, status, alarm log entry number, and space for an operator entered text area. The operator can be given wide ranging ability to print, display, or update alarm log entries at any time.

Analogous to the alarm tracking mechanism, the NCC can provide a computerized trouble ticket log to coordinate and track general troubles. The trouble ticket format can provide time, date and trouble ticket number entered by the system. In addition, there can be fields for the operator's name, the site telephone number, an area for trouble descriptions and a status field (open and closed).

Both alarm and trouble ticket management summary reports can provide management with visibility, as well as documentation of problems handled by the staff, and may be used as a gauge of transmission performance.

5.7 SPECIALIZED TEST EQUIPMENT

External diagnostics or portable data communications testers have decided advantages over central testing and internal diagnostics. First, onsite testers do not burden the network with the cost of building the diagnostics into the network or tying up valuable computer time to run them. Second, now that electronic components are denser and more powerful, data communications testers are lighter, more compact, and more powerful as well.

With added flexibility, onsite testers can be programmed to fit nearly any network situation: superior testers are able to run bit error rate or protocol tests; monitor the RS–232 interface; and simulate both data terminal equipment and data communications equipment, checking the signal coming from both directions.

A network is composed of many elements, including a mainframe, a front end, modems and carrier links, just for hardware. Then there is the software. Each of these network elements may come from a different vendor as illustrated in Fig. 5–4. The complexity of today's networks leads to finger pointing. The user points to the most highly suspect vendor or carrier; the vendor to a more highly suspect vendor, to the carrier, or back to the user. When the system goes down, who should be called?

At times like these portable data communication testers are very useful. A user can check out the vendor's equipment handshaking capability by using a data communication tester to simulate the protocol that the network will use. Similarly, vendors can assure users that the equipment is operational, even though there is no communication link at the site.

In more complex networks that use bit-oriented protocols at higher levels, the portable testers can be modified to troubleshoot links at those levels. The same testers can be modified to handle simple character-oriented, asynchronous protocols at lower network levels. This same flexibility makes network testers useful.

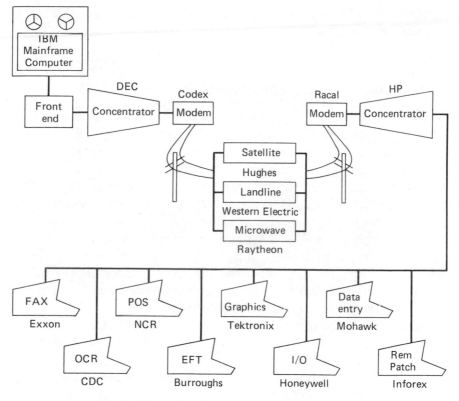

Fig. 5-4. Typical multivendor communication network.

5.7.1 Line Analyzers

Line impairment tests are performed using analog measuring equipment and are conducted mostly on telephone lines. Tests of this nature measure all types of line characteristics that the reader may or may not be familiar with (e.g., frequency response, signal/noise ratio, envelope delay distortion, frequency translation, impulse noise, etc.). The hardware required for these tests is sophisticated (Fig. 5-5) and the operator must be skilled at interpreting the results. This hardware should only be required at a user's site when extremely troublesome lines are encountered and proof is needed to convince the carrier that action need be taken. Usually, digital instrumenta-

Fig. 5–5. Hewlett-Packard 4935A transmission impairment measuring set.

tion, while not providing data accurate enough to pinpoint the problem, is sufficient to identify that there is a problem. If some audio capability is required, a low-cost line monitor with a speaker, test tone generator, and power level meter will usually suffice.

5.7.2 Breakout Boxes

These devices are necessary in troubleshooting and in identifying interface/timing incompatibilities. This capability is part of the majority of more sophisticated devices but the breakout box can be purchased separately. It usually consists of two connectors (one for connection to the DTE and one for the connection to the DCE. The device is connected in series between the two elements with the status of the various leads visible through indicators (usually light emitting diodes, LEDs). Some of the units use tri-state LEDs (red, green, off) to show if the interface lead is high, low, or off. In addition, the leads are generally brought out to pinjacks to allow connection to external test equipment. Most of these are non-interferring with data and control signal flow between the DTE and DCE.

5.7.3 Error Rate Testers

The two most commonly used devices of this type are bit error rate testers (BERT) and block error rate testers (BLERT). They usually generate a known bit or block pattern. With most that are currently available, user selected test data can be generated as well as standard

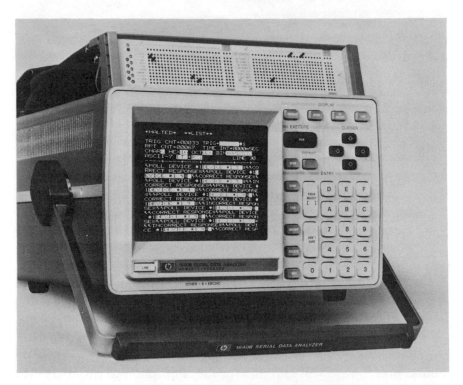

Fig. 5–6. Hewlett-Packard 1640B serial data analyzer.

"fox" messages and pseudorandom patterns. The device monitors what is returned to it and provides a numerical readout of the BER over a specified time frame. This is a good overall test of the transmission facility and hardware. It gives a reasonable indication that the problem is random, solid, or occurs periodically in bursts.

5.7.4 Data Monitors

The function of these units is to bridge the connection between a DTE and DCE on a non-interferring basis and to display data and events (Fig. 5–6) as the connected devices interact. The display is usually a CRT and an optional printer port is sometimes provided (not the printer). These units always have internal storage and many are equipped with an integral tape unit, or the means to connect to an external unit, in addition to semiconductor memory.

5.8 NETWORK SECURITY

Security vulnerabilties increase as the communication link gets longer. The communications links within a building are the easiest to control because, depending on the distance between each communicating device, they can be totally organization owned and inspected. The most common transmission media include twisted wire, coaxial cable, and fiber optic lines. The intrabuilding link security problem increases when the organization leases lines or uses voice dial-up circuits. Here the security problems become similar to those of the local area category because the organization has relinquished control.

The local area network may be entirely within a large building, an industrial complex, or a small area of city. Its communications links can be user-owned cables (including fiber optics), microwave circuits, leased lines, and voice dial-up. The threat of data alteration via line taps is greater than in intrabuilding. Line taps are easy to install and and difficult to prevent.

Domestic networks via commercial carrier, including value-added networks, greatly increase link security problems. Unless private lines are used, data are transported totally by vendor media. This may be a plus where packet-switching networks route various pieces of information via different circuits, but it is a drawback where satellite transmission is concerned. Satellites provide the least degree of data security because transmissions may be intercepted by anyone in the area with an appropriate antenna.

International networks have all the security vulnerability of domestic networks with two additional problems: they are subject to transborder dataflow limitations and, sometimes, data encryption is not allowed. In addition, many foreign countries have poorer-quality communications networks than those available in the U.S.

Data encryption (Fig. 5–7) is the process of changing original data (cleartext) into ciphertext so that they cannot be understood until the ciphertext is decrypted into cleartext at this distant end of a link. The NBS formal description of the data encryption standard (DES) is the DES algorithm—a recirculating 64-bit block whose security is based on a secret key. DES keys are 64-bit binary vectors consisting of 56 independent information bits and eight parity bits. The 56 information bits are used by the enciphering and deciphering operations and are referred to as the key. Active keys are selected at random from all possible keys by each group of authorized users. Each user should understand that the key must be protected and that any compromise of the key will compromise all data protected by that key.

In the encryption computation the 64-bit input is divided into two halves each consisting of 32 bits. One half is used as input to a com-

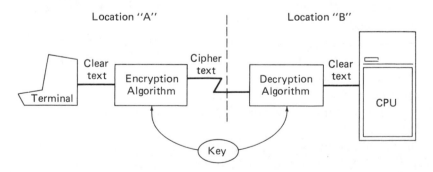

Fig. 5–7. Conventional cryptographic system.

plex nonlinear function, and the result is exclusive OR'ed to the other half. After one round, the two halves of data are swapped and the operation is performed again. The DES algorithm uses 16 rounds to produce a recirculating block cipher.

Encryption, however complex the process may seem, is the only realistic way of protecting data in transit. Few security managers would argue the need for system protection, but convincing management acceptance due to costs and system resource implications is the real problem to the solved, prior to installation of any cryptographic products. Management must be prepared to weigh the costs and benefits of protecting the organization's information system to decide whether encryption is worth the expense involved. Since some cryptographic systems use cpu cycles, they affect the overall communication system performance and response time. The degree to which performance is affected depends upon the specific hardware and software used. Consequently, the project manager should demand that vendors demonstrate exactly what effect their product will have on the operating system—prior to purchase and installation.

6
NETWORK STRUCTURES

6.1 TOPOLOGY TRADEOFFS

If we look at an overall network we see a certain architecture. This architecture is called the topology. It defines the way in which the work stations and peripheral support devices are interconnected to form a network.

Networks can be broken down into a few basic types. Star configurations (Fig. 6–1) have the work stations arranged around a central hub with radial links going between the hub and the work stations. In this network all communication between the work stations must pass through the hub. This can provide fast response time to communication requests. However, it is more susceptible to sys-

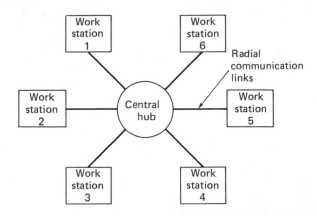

Fig. 6–1. A star configuration with a single dedicated hub.

tem failures due to the single centralized communication hub. This configuration finds its best use in networks which must communicate over long distances. The star configuration is an example of a non-contention network. In other words, work stations wishing to communicate with others don't have to contend for a communication link as compared to a contention network, of which loop technology is an example.

Loop topologies or rings are typified by work stations arranged in a ring with a communication link going from one station to the next. This topology requires the work stations to cooperate in their use of the communication link. This cooperation can be in the form of master/slave allocation schemes or token passing. Essentially, the allocation of the link must be accomplished without interfering with a station already using the link.

Figure 6-2 shows an example of a multidrop master/slave topology. The master originates all communication with the slave responding at the appropriate time. The slaves serve the role of sensors for the master. The system finds its weakness when the number of sensory inputs gets larger than about 25 devices. The problem is that the slave can only transfer information when asked by the master. As the number of slaves goes up the response time goes down.

Figure 6-3 is an example of a token passing ring topology. Token passing is a line access method, where control of the rights to access are distributed among all devices using the network. The only device that has the right to transmit is the device holding the "token" or

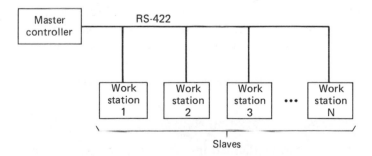

Fig. 6-2. A typical polled multidrop master/slave configuration.

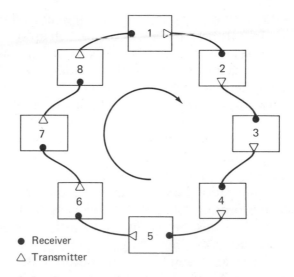

Fig. 6-3. In token-passing ring topology, access rights are distributed.

message. After the device has finished its transmission, it passes the token to another device. The token is passed from device to device in a logical sequence.

Token passing is attractive because the transmission delay can be calculated for any network. This makes it ideal for use in real-time applications. The big drawback of the token passing method is that it is more complicated. Recovery circuits have to be built into each device to ensure that a sudden power loss will start the token message again.

Another common type of broadcast topology often used in LAN networks is the bus. A bus network (Fig. 6-4) consists of a linear length of cable to which user stations are connected by cable interfaces, or taps. A single cable can carry messages from each tap in both directions on either one channel, or two separate channels.

6.1.1 Front-End and Back-End Networks

Data networks can be divided into two distinct groups based on the application requirements. Figure 6-5 shows a system with two CPUs, several users and shared storage resources. The user interface to the

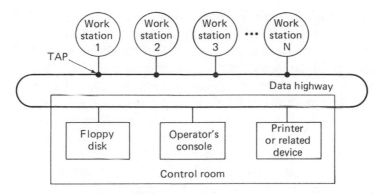

Fig. 6-4. A bus network, typical configuration.

processors is the frontend network, and the network to which the storage and input/output devices are attached is the backend network. Although a linear bus interconnection is shown in the figure, these networks can have any topology. Frontend networks are characterized by a large number of devices such as word processors, CRT terminals, printers and copiers.

Backend networks serve computer-to-computer communication and computer-to-hard disks or high-speed device communication.

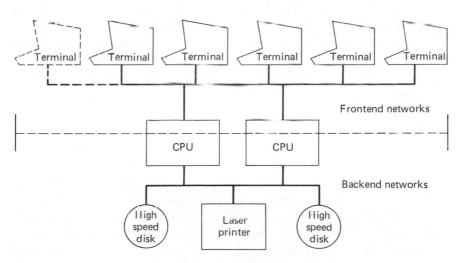

Fig. 6-5. Frontend and backend networks.

Backend networks typically have tens of nodes which need to communicate at very high data rates in the 50 to 100 MBS range. Frontend networks can have hundreds of nodes and because of the human interface involved in all these, data rates of a magnitude lower can be adequate in many cases. Also backend networks have the nodes concentrated in shorter distances in the order of a kilometer, than frontend networks. Both bus and ring topologies are used in backend networks and rings can provide an upgrade to even higher speeds using fiber-optic cable as the medium.

Frontend and backend networks also differ in terms of the implementation cost. Frontend devices are usually low performance, low cost devices compared to backend devices which are quite expensive. Hence backend networks can incorporate more sophisticated features and hardware to obtain the performance they need.

6.2 LOCAL AREA NETWORKS

The diversity of commercial local area networks (LANs) is one of the strongest considerations of the current marketplace. Each LAN product offers unique components, functions and applications. When a local area network is classified by topology, it is the access method of each product that sets it apart from its competitors. The access method is the set of procedures by which a user device is granted access to a data channel. General characteristics of an access method include distribution of the right to communicate on a fair and orderly basis. The right to access can be controlled by a central arbiter or shared by all stations in the network. Assignment of special priorities or different levels of access may or may not be accommodated.

The access method also determines the time a station may have to wait before gaining access to the network. Some access methods like token passing are "deterministic," meaning that the delay time is a predictable fixed function dependent on a particular network specification, such as number of network nodes. Others, are "statistical," meaning that only the odds of gaining access are calculable, not the exact waiting time itself. Deterministic networks are easier to model than statistical networks, when the effect of a change involving data rates, message size, or other variables must be determined.

The network's ability to recover from the failure of a critical network component, such as a node, is also determined by the access protocol. The access method also determines the stability with which the network accommodates exception conditions, such as operating under extreme loads, in terms of maintaining acceptable delay times and fair access.

There are nearly as many types of access methods, as there are LAN network offerings. The types can be broken down into two basic classes: centralized access methods, in which a central arbiter controls the user stations' right to access; and distributed access methods, in which the individual user stations control the right to access.

6.2.1 The Ethernet Way

Local Area Network interconnection schemes such as the Ethernet provide the framework in which one can construct a system of sharing resources in an effective manner. Ethernet uses a broadcast mechanism (coaxial cable) and a distributed access procedure to allow for sharing of the channel. The procedure is called carrier sense, multiple access with collision detection (CSMA/CD). Nodes on Ethernet can sense ongoing transmissions and defer theirs until the channel is idle. They also have the ability to monitor the channel while transmitting to determine if any other stations are also attempting to transmit. Once an idle channel is sensed a station may transmit. Because of the propagation delay on the wire, two or more stations may sense an idle channel and attempt to transmit simultaneously. This results in a collision.

In order that all stations (including the one transmitting the packet) can "hear" the collision it is required that all packets be greater than a certain minimum size. That size is determined by a parameter called the "slot time". The slot time is slightly greater than the round trip propagation delay. Any station involved in a collision must stop sending the packet and reschedule the transmission. The algorithm used to determine when the next attempt should be made is called the backoff algorithm. Basically, every time a station is involved in a collision it backs off (i.e., waits) a random amount of time that increases every time it experiences a collision. The backoff time is reset after a successful transmission.

One of the main reasons for Ethernet's popularity is because it uses a passive broadcast medium. This results in very reliable operation. Ethernet inferfaces can be built using VLSI technology and thus made fairly inexpensive. Multivendor environments can be implemented by adhering to interface specifications at any of several levels. For instance, one may choose to provide compatibility at the wire tap, transceiver cable or port.

Ethernet is a shared bus network topology. The shared physical coaxial cable is passive and therefore highly reliable. The key active component which interfaces nodes is the transceiver. Ethernet is all-digital; there are no analog-to-digital conversions nor modems needed for transmission of data on the coaxial cable. Maintenance of the network and repair of faulty components can be accomplished without interrupting network operation. For example, any node can be powered off, repaired, and then powered back on again, without causing a disruption of service.

Coaxial cable, in conjunction with distributed CSMA/CD network control, enables communications at ten million bits per second, in order to accommodate the high-speed, bursty transmissions of a large number of participating nodes. Up to 1024 nodes can connect to Ethernet, and each node can accommodate a number of devices.

A unique characteristic of Ethernet is that messages are broadcast over the physical medium. That is, once the signal is placed on the cable, it travels in both directions, away from the sending node and out over the entire network. All nodes hear the message as it passes by, and each of them checks to see if it is addressed to itself. However, observation of network traffic by the nodes is passive. They do not have to handle passing messages addressed to other nodes.

Compatibility at the level of the transceiver cable interface in a network, allows nodes to be connected at wall outlets. The transceiver cable can be dropped from the coaxial cable (Fig. 6–6), that is, run through the walls or over the ceiling, and can terminate in a wall outlet for network interfaces. Connection to the coaxial cable

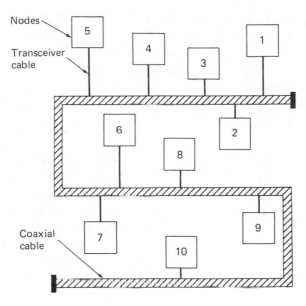

Fig. 6-6. A typical ethernet configuration.

is achieved by a simple tap. In both cases, network operation is not interrupted. The problem of managing thousands of wires is greatly reduced and there is no central switching or complex routing procedures required for communication.

6.2.2 LAN Cable Medium Choices

The process of putting a local area network together ultimately includes the selection of a suitable cabling medium. After such issues as architecture and topology have been addressed, it is necessary to evaluate cable technologies in light of these issues and their requirements. In addition, user concerns such as cable size, installation ease, costs, and reliability must be included in the planning process.

For the media requirements of a local area network, both optical fiber and copper cabling should be considered. Each has strengths and weaknesses. For example, while optical fibers offer complete immunity to electromagnetic interference, copper wire is lower in cost.

The major advantage to using copper at data rates between 1 and 10 MBS is that it involves a mature, well-understood technology. Connectors are readily available, splices are simple, and both the production of cable and the installation techniques required are well known to manufacturers and users. The major disadvantages of copper wire, as compared with optical fiber, are that signal attenuation on copper increases with frequency, while attenuation on an optical fiber is constant up into the 300-MHz region (Fig. 6–7). Copperbased networks are therefore more expensive at high frequencies. Also, copper cable requires a surrounding shield to reduce noise pickup from the environment, as well as to reduce electromagnetic radiation from the copper itself. In light of recent Federal Communications Commission outward-radiation rules, this presents a design problem whose solution requires grounding the shield at both ends of the cable. But this introduces another potential problem: For long cable runs, grounding both ends of a shield can introduce ground currents if the ground voltages at the end points are not equal. Optical fiber is completely immune to such problems.

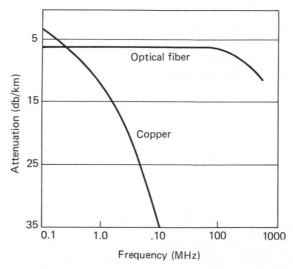

Fig. 6–7. Copper versus fiber. Signal attenuation on the copper medium increases with frequency, while on optical fiber it is constant up into the 300-MHz region.

In contrast, the drawbacks to optical fiber are the lack of a low-loss optical-fiber connector that can be easily assembled and installed in the field. The splicing techniques also suffer from similar limitations. Also, for low-bandwidth applications (below 10 MHz), optical fiber is more expensive than copper. This is because optical-fiber cable and components are still manufactured in relatively low volumes. In addition, it is more difficult to find quality installation and maintenance expertise for optical fiber than for copper cable.

Disadvantages of optical fiber are transitory, since the technology is evolving rapidly. The cost of optical-fiber cables is expected to come down; better connectors will be developed; and more operating expertise will be acquired. In spite of the many advantages of optical-fiber technology, there remains a place for copper media, especially when the data rates are moderate (10 MBs or less).

If the decision has been made to use copper in a local area network, the question of twisted-pair versus coaxial cable must be addressed. Coax is an unbalanced medium—the shield and the center conductor have different impedances to ground. Twisted pairs, in contrast, are normally balanced; both wires have the same impedance to ground. Since coaxial cable is unbalanced, it is subject to spurious signals that may be picked up in both wires of a pair. The source of this noise is generally outside the cable. If the cable consists of a well balanced twisted pair, this noise affects both wires equally. If the wire pair is terminated in a balanced receiver the noise is generally eliminated.

To keep electromagnetic radiation from affecting the transmission medium, a grounded shield must surround the wires. Since the shield (in the case of coaxial cable) is part of the signal circuit, grounding it introduces noise. A second shield, as in multi-shielded coax, alleviates this problem, but at great expense. However, coaxial cable generally has lower attenuation than twisted pairs, especially at higher frequencies.

6.3 DISTRIBUTED COMMUNICATIONS

There are two generic ways that applications can be distributed so as to minimize communications and allow for the balancing of processing among several nodes in a network. These two techniques can be used in combination in much the same way that series and parallel connections of electrical components can be used to implement any electrical circuit.

Horizontal distribution is typically characterized by each of the nodes performing essentially the same functions. Usually the differences between the functions are in the data stored on each of the nodes and the terminals to which they are connected. Communications between the systems take the form of program-to-program requests and responses in both directions. These communications are minimized due to existing relationships among the terminals connected to a particular system and the data stored there.

As an example, consider two nodes used for retail credit authorization. Assume that each node is assigned to cover a particular geographical area. Terminals for each area are connected to their corresponding node, along with the stored data that reflect the account status of the card holders for that area. The level of communications between the two nodes would be a function of probability that a card holder from one geographical area is shopping in a retail establishment in the other node's geographical area.

The state of California is an example in which the geography is ideal for this type of split because the major population centers, San Francisco and Los Angeles, are separated by a large zone with relatively little population. Therefore, it would be relatively easy to balance the load between two nodes with the probability of communications between them being quite low. In contrast, in a single-center geographical region such as New York City, this approach would not be practical.

In a vertical distribution arrangement, the terminals are usually placed on the smaller node, with another node being placed upstream. In such a configuration, data for the relatively simple functions would be maintained on the nodes that are connected to the terminals. Further, the application would be designed so that the bulk of the transactions that originate in these terminals would be handled by the local system. Only in the event of a more complex request, or one that requires data stored on the upstream node, is a request sent to the remote node. Thus the form of the communications in a vertically distributed system is that of requests flowing in one direction and responses in the other.

An example of vertical distribution is an application wherein the inventory of a local warehouse is maintained in a machine at that site. If, during order entry, a particular item is not in stock locally, an inquiry might be sent automatically either to the machine that is servicing another warehouse or to a central location to determine whether the item is available elsewhere.

6.4 MESSAGE SWITCHING NETWORKS

Message switching has been in operation since the earliest communications systems were installed. The initial systems, called "torn tape" systems, used paper tape to capture traffic. Operators would then read the message address and manually take the tape to the proper machine for retransmission. Since then dramatic changes have occured in message switching, affecting the size and cost of the required equipment.

Microprocessor technology has now made it possible to buy a computer based switch capable of handling a medium-to-large company's traffic for less than $50,000. In general, this approach offers significant cost advantages as well as security and control which many companies desire when transmitting sensitive intra-company information. Many companies are therefore reassessing their communications strategics with a view towards centering their operations around an in-house message switch.

Message switching is a form of data communications where messages are sent along with an address to a central station which then forwards the message to the intended destination. Its major difference from most other forms of data communications is that originator and recipient are not both on-line simultaneously and no two-way communication is possible. Alternatives to message switching include leased-line networks, circuit switching, or some form of manual delivery (see Fig. 6–8).

If communication is needed between only a small number of sites, a communication network can be built linking each of those sites with a fixed line. These lines could be terminated with phone sets or teletypewriters. This setup is attractive in two ways. First, communication takes place in real time, allowing a dialogue, where desirable, between the stations. Second, any station can communicate with

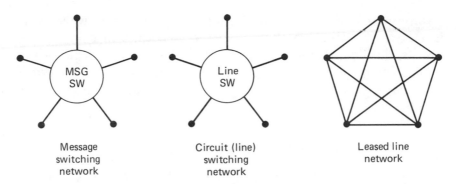

Fig. 6–8. Communications network configuration alternatives.

any other station at any time and there will never be a "busy" when trying to deliver a message. As the number of sites increases, however, the number of required lines increases geometrically to the point that leased line cost quickly becomes prohibitive.

Circuit switching is much more economical in line use. In this method, a line is placed between each site and some central station. Therefore, for n stations only n lines are needed. Connections are then made by command at the central station to allow communications between sites. Like leased line networks, all communication takes place in real-time. A connection, however, is not always possible since there is only one link from the central station to each site.

In message switching, the problem of "busies" encountered in circuit switching networks is eliminated, since traffic is simply sent to the central site and captured there. The switch will then hold the traffic until the destination leg is able to receive it and will then forward the message (Fig. 6–9). The only feature lost is the real-time connection allowing conversational communication. However, this is infrequently used in data communications and its lack is generally not a serious drawback. This is especially true in international communications since the originator and recipient of a message are seldom working at the same time due to the time-zone differences. Line costs can be reduced even further than for circuit switching networks. This is because traffic loads can be distributed over a longer period of time. Peaks in traffic are buffered at the switch and sent out steadily over the destination lines. This allows either fewer lines

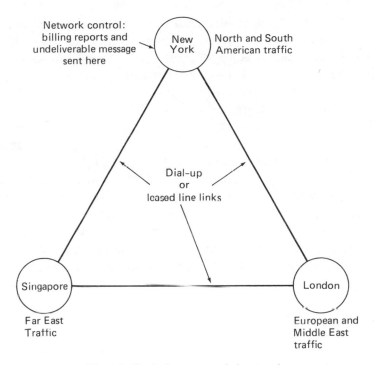

Fig. 6-9. Typical message-switch network.

to each site, or lines of a lower speed to be used to handle the same traffic loads.

Perhaps the most powerful and useful capability found in message switching is the ability to access a variety of different services. A good message switch can control leased lines in addition to interfacing to the Telex, TWX, DDD, international Telex, and international DDD networks. Users can therefore access any of these services through proper addressing of their messages. This capability makes it possible for private networks to communicate compatibly with existing public networks.

6.5 INTEGRATED VOICE AND DATA

It is wise to plan for the future. Even the corporation that has no identified near-term need should ensure that its telecommunication network can accommodate high speed data switching and transmission. The present level of office automation already demands efficient, low-cost means of interconnecting terminals at the local level. Electronic mail, high quality facsimile, and teleconferencing are examples of emerging applications that require long-distance switched data communication between local areas.

The reasons for integrating voice and data are straightforward. Sharing switch capacity and transmission bandwidth significantly reduces cost. A single 56 kbps voice channel, for example, can carry the multiplexed output of tens of 1200 bps data terminals. The ultimate design objectives for voice and data networks are the same: timely delivery of an accurate replica of the transmitted message at minimum cost. Implementation techniques differ to the extent that they derive from different backgrounds—telephone and computers. Integration of digital voice and data, on the other hand, takes advantage of the best of both and yields a network of least cost to the user.

6.6 SHARED RESOURCES

Communication networks grew, primarily, from a need to make the resources of a computer-based system available to a geographically dispersed community of users. The earliest networks were simple circuit-switched channels interconnecting the computer to the users on a point-to-point basis. As private networks grew, it became more evident that sporadic use of dedicated circuits often could not be economically justified. The user was being allocated costs based on time and distance. Circuit idle time was often high. A better solution was needed—one which would allow users to share the cost of expensive communication facilities. By the mid-1960's, technology had progressed to the point where the first networks based on the concepts of shared use of the communication facility among users became practical (see Fig. 6–10).

The shared use of communication facilities by an entire community of users led naturally to the consideration of providing these facilities as a public rather than a private resource. From this arose the development of the so-called public data networks (PDN) (see Fig. 6–11). Public data networks allow users to share expensive com-

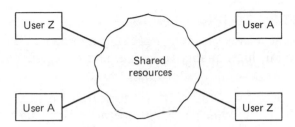

Fig. 6–10. Shared community resources.

munication facilities. The user, called a subscriber in PDN termi-
nology, has dedicated access only to the first node of the network,
called a data switching exchange (DSE). The rest of the facilities are
shared among all subscribers on an as-required demand basis.

Public data networks are spreading rapidly around the globe. Most
countries now offer this type of service. In the United States the two
major public networks are operated by GTE Telenet and TYMNET.
Other specialized public networks are in operation and more are on
the horizon.

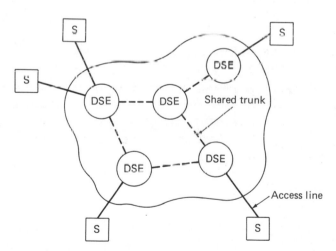

S: Subscriber
DSE: Data switching exchange

Fig. 6–11. Public data network.

6.7 PACKET SWITCHING NETWORKS

The pioneering efforts towards sharing the resources of networks among a community of users were based on a technique of channel allocation called packet switching. The first successful packet switched network was the Advance Research Projects Agency Network (ARPANET) which became operational in the United States in 1969, followed soon after by others in Europe.

Packet switching is a derivative of message switching and uses many message switching features. The key difference is that the message is divided into maximal length blocks called packets. The channel capacity is allocated to the user only for the duration of each packet. Each packet is individually addressed and transmits the network in a store and forward fashion (see Fig. 6–12). When a packet is transmitted, the channel is immediately available for allocation to packets being generated by other users. In this way the expensive facility costs may be shared by many users.

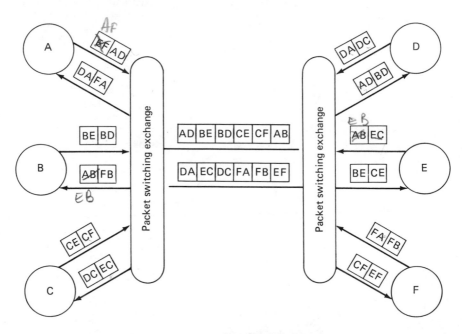

Fig. 6–12. Packet switching addressing scheme.

Packet switching networks use the concept of virtual calls and virtual circuits. When a request to enter packets into the network is received a 'virtual call' is established. The virtual call is somewhat analogous to a dialed switched network connection (see Fig. 6–13). It exists only for the duration of the call. For that period the network behaves as though a fixed path exists from the source to destination. Those users with a high volume of traffic can establish a permanent virtual circuit. This type of path behaves like a dedicated channel and eliminates the need for set up and termination sequences.

The virtual circuit concept is made to work by the use of logical channel numbers. Each packet entering the network is assigned a logical channel number which indicates the session or 'conversation' to which it belongs. The network can then associate the logical channel number of one subscriber to the logical channel number of another.

6.7.1 The History of X.25

As packet switching became feasible, an obvious need for a common protocol which would allow users to interface a PDN became evi-

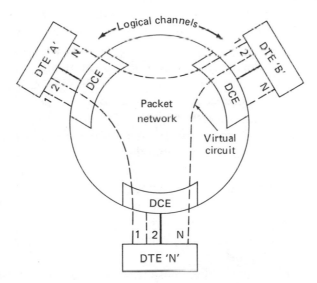

Fig. 6–13. Virtual circuit service concept.

dent. Various national and international standards groups began working on the problem in 1972. In 1976, the CCITT formally adopted recommendation X.25 with the imposing title: "Interface Between Data Terminal Equipment and Data Circuit Terminating Equipment for Terminals Operating in the Packet Mode on Public Data Networks". The recommendation was revised in 1980 to add additional features and enhancements.

CCITT X.25 defines the protocol and procedures necessary for a packet mode terminal to access the services provided by a packet-switched public data network. The services and the related facilities provided include virtual and permanent virtual circuit service. A packet mode terminal generally operates synchronously over full-duplex circuits at speeds of up to 56,000 bits per second.

6.7.2 How They Work

The current interest in the X.25 standard and packet switching has left many people with less than perfect grasp of the concepts involved. The concepts are not difficult to understand, but unless stated clearly, many of them easily become confused.

A packet switching network provides what appears to be a permanent connection between devices. On the functional level, a packet network handles blocks of data the way the Post Office passes an envelope through the mail. That is, none of the facilities is dedicated to any one message. The network facilities handle many different messages in quick succession.

When the terminal requests a connection to a computer port via a packet switched network, the form of the request is a packet or data block. It contains specific control bits that identify it as a call request, the address of the called port, the address of the calling station, and some additional overhead information. This packet passes from node to node within the packet switching network (Fig. 6–14) until it reaches the called port. The device there acknowledges the call request with a second specialized form of packet that is returned to the calling device.

As these two special packets pass through the network, the nodes that handle them assign a serial number to the call. This is simply a

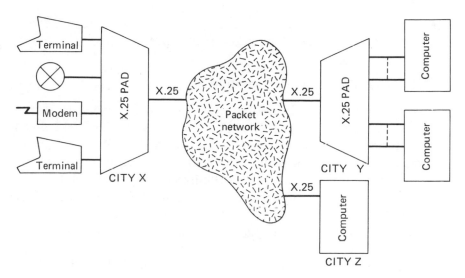

Fig. 6–14. Packet switching network connections.

large number that is unique to the access link while the call is connected; after disconnect, the number is available for use again. The "logical channel number" is listed in temporary memory at each node within the calling station, the called station, and the appropriate routing (which data links to use).

The end-to-end route determined by this logical channel number is the "virtual circuit" between the devices. Each device needs only to include the logical channel number in the packets to be routed to the correct destination. Packet switching networks thus offer the conveniences of variable routes without the need to create a separate private network. Each user "plugs in" a terminal or computer without concern for the nature of the network or even what locations he passes through.

What the user "plugs in" is the packet assembler/disassembler. The PAD sits between the terminal and the network. The PAD stores input from a typist in a terminal buffer, of say 256 bytes. The PAD then will assemble and create a packet by emptying the buffer when it is full (after almost a minute, at the average rate of five characters

per second of a typist). For faster response a PAD can be programmed to send a smaller packet after a present time.

6.7.3 Packet Switching in Operation

Packets move around the network, from switching center to switching center, on a hold-and-forward basis. That is, each switching center, after receiving a packet, "holds" a copy of it in temporary
storage until the switch is sure that it has been received properly by
the next switch or by the end user. Unlike message switching, which
uses the principle of store-and-forward switching, the copies of the
packets are destroyed (actually, written over in memory) when the
switch is confident that the packet has been successfully relayed.

The flow of the message, in Fig. 6–15, is initiated by the transmission of packet 1 between user A and switch 1. When it fully receives the first packet, switch 1, following a set of routing rules,
transmits packet 1 toward its destination by sending it via switch 2.
In the meantime, packet 2 is moving from user A into switch 1. During this time the conditions in the network change (for instance, a
large amount of traffic from switch 5 arrives at switch 2), so the
second packet of the message from A to B is routed via switch 4.
The third packet of the message, arriving at switch 1 soon after the
second packet, is similarly routed via switch 4.

After being received correctly by switch 4, the second packet is
transmitted to the destination switch, switch 3. But during that transmission an error occurs. When switch 3 receives packet 2, the error-
detection mechanism is able to detect the error and request a retransmission of packet 2. However, while this is occurring, packet 3 has
been transmitted immediately behind the first (and errored) copy of
packet 2. As a result, the second (correct) copy of packet 2 is received
at switch 3 after packet 3. If we look at the network from the perspective of switch 3, first packet 1 is received, then packet 3, and
finally packet 2. If switch 3 delivered the packets to the destination
(that is, to user B) in the same order they arrived at switch 3, user
B would receive the packets in a different order than they entered
the network.

Differential delays along the many paths through the network in-

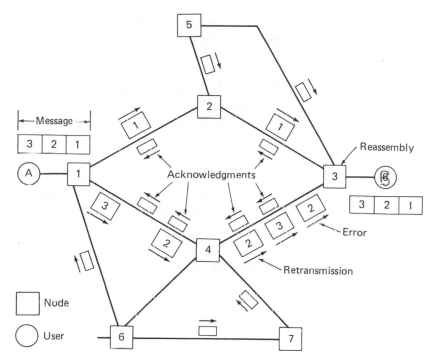

Fig. 6–15. Packetized movement of messages in a network.

troduce the possibility that packets will be received out of proper
sequence. For users to be protected from this form of error, the
packets have to be reassembled into the same basic message structure
they had upon initial transmission into the network. The process of
packet reassembly is done at the destination switch—in this case,
switch 3-using packet sequence numbers that must be carried through
the network.

6.8 NETWORK DESIGN CONSIDERATIONS

Network design requirements can be divided into two categories:

1. *Existing facilities.* These can be divided into the following items:
   • Number of terminals supported

- Speed/code/protocol of each terminal
- Physical location of each terminal
- Physical location of the host processor
- Speed/code/protocol of the host processor

2. *Traffic statistics*. These can be divided into the following items:
   - Average number of terminals in use at any one time
   - Average length of input inquires
   - Average length of output responses
   - Peak hours in different times zones

Once the basic traffic requirements (Table 6-1) are known, the next task is to organize the data. The first part of this organization

Table 6-1. Network Traffic Statistics

a) Number of terminals supported... 440

b) Speed/Code/Protocol of each terminal... •300 baud •ASCII •ASYNCH.

c) Average number of terminals in use at one time 50%/200

d) Average number of inquires input from a terminal per hour 240

e) Average length of input inquires .... 6 characters

f) Average percent of time terminal is in input 78%

g) Average number of responses output to a terminal from the Host per hour 240

h) Average length of output responses... 37

i) Average percent of time terminal is in output 22%

j) Physical location of terminals:

Location	ATL	CHI	BOS	NYC	MIA			
Number of Terminals	120	120	40	120	40			
Number Active	60	60	20	60	20			

k) Physical location of host .... NYC

l) Speed/Code/Protocol of host... X.25

is to construct a logical to/from matrix. This matrix indicates the possible source-to-destination connection. Once the logical matrix is constructed, several conclusions can be drawn that will indicate a possible network configuration (see Fig. 6–16).

The next step is to determine the topology of the backbone network. The topology is determined by three basic considerations.

1. *Physical considerations.* This includes existing or planned facilities and geographical preferences as well as considerations such as access to communications facilities.
2. *Cost considerations.* This includes the physical facilities, labor and communications costs.
3. *Requirements of the network.* This would include the pattern and volume of the network traffic as well as considerations of redundancy and network growth.

The decision on backbone topology is the most important and often the most difficult part of network design because of physical

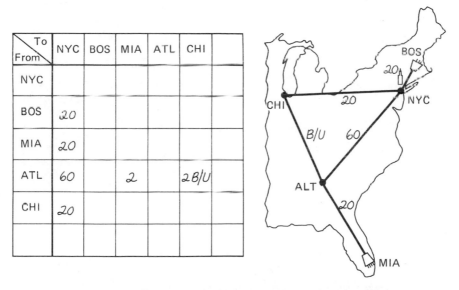

To From	NYC	BOS	MIA	ATL	CHI	
NYC						
BOS	20					
MIA	20					
ATL	60		2		2 B/U	
CHI	20					

Fig. 6–16. Logical matrix. Each square is filled in with the number of connections required between intersecting cities.

and cost considerations. If traffic patterns and volume were the only considerations, then network topology could be generated by computer software. This would lead to the best network solution. However, physical and cost considerations play an important part in the design, forcing the use of a combination of techniques to arrive at a solution. This means that given all existing considerations, the final network design is only a compromise.

Configuring redundancy to back-up lines and equipment failures always involves a tradeoff between availability and cost. Even minor changes to the redundancy philosophy can have a major impact on network cost. Redundancy requirements within the network are usually assessed from three perspectives.

1. *Node hardware redundancy*. This includes redundancy of the processing hardware and communication port hardware.
2. *Trunkline redundancy*. This includes alternate routing of trunkline traffic around backbone failure.
3. *Netline redundancy*. This takes the form of alternative access paths to the backbone nodes.

7
SATELLITE AND
CARRIER SERVICES

7.1 SATELLITE OVERVIEW

At an altitude of 36,000 kilometers above the equator, telecommunications satellites are by no means isolated from the chain of ground-based networks, of which they are a vital link. The role of satellites is growing steadily. Their altitude above the equator enables them to cover a full one-third of our globe. The possibilities they offer are also expanding because of their growing transmission capabilities and the anticipated development of intersatellite links. The latter will provide direct communication between geostationary satellites (domestic and international) or between geostationary satellites and data-relay satellite systems, thus setting up other communication networks in space.

Figure 7-1 shows a simplified satellite channel and its components. The channel starts at the host computer, which is connected by way of a local communications loop (using traditional telephone company facilities) to the central office of the satellite communications vendor. The data from the local loop is then multiplexed with other data received from separate sources into a microwave signal that is sent to the satellite vendor's earth station. This signal becomes part of a composite signal that is sent by the earth station to the satellite and is transmitted by the satellite to the receiving earth stations. The satellite uses a transponder (a device that receives RF signals at one frequency and converts them to another frequency for transmission)

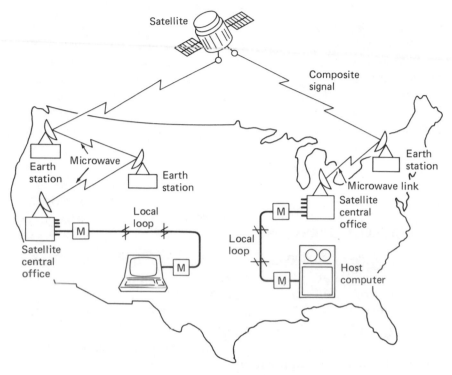

Fig. 7-1. Satellite channel components.

to transfer the composite signal from one earth station to another. At the receiving earth station, the data is transferred by a microwave link to the satellite vendor's central office; from there, it is broken down into separate communications channels and transmitted over local telephone company facilities to the receiving terminal.

The distance between the host computer and the receiving terminal is essentially the same for all points within the area serviced by the satellite, since the satellite is about 23,000 miles above both the transmitting and receiving earth stations. The major costs in leasing a satellite circuit are those of the terrestrial stations, the central office facilities, and the satellite link itself. The cost of a satellite channel connecting, say, Washington, D.C. and Atlanta, GA, is equivalent to the cost of a satellite channel connecting Washington, D.C. and Los Angeles, CA.

The composite signal between the satellite earth stations and the satellite is assigned to a frequency in the billion-cycle-per-second range (gigahertz). The amount of information transferred over a communications channel is directly proportional to the available frequency bandwidth of the channel; therefore, satellite channels operating in the high end of the frequency spectrum have large transmission bandwidths and can easily send and receive data up to millions of bits per second.

Satellite channels also provide low error rates. The composite signal from the terrestrial stations runs up out of the earth's atmosphere directly to the satellite and down through the earth's atmosphere from the satellite to the receiving earth station. Although normal terrestrial communication channels use similar microwave circuits for long-haul transmission, these circuits run through the lower atmosphere, more or less parallel to the ground. The terrestrial microwave circuits are therefore somewhat more subject to atmospheric interference than are the satellite channels.

7.1.1 Overcoming Attenuation

Only a limited number of communications satellites are in synchronous earth orbit, covering specific geographic areas within a pair of radio frequencies. Simultaneous transmissions at the same frequency in the same place interfere with each other. Sufficient spacing between satellites must be provided to ensure a low level of interference from earth stations and adjacent satellites.

In an up-link (from earth station to satellite), a loss of signal power due to rain can be overcome by supplying increased power to the transmission. Rain attenuation of the signal is significant only when there are heavy thundershowers. Fortunately such thundershowers usually cover only a small area of about a mile in diameter. One method of overcoming rain attenuation is to transmit data to alternate earth stations, carrying it over interconnecting facilities to the appropriate distribution points. A less costly method is to provide more sophisticated transmission technology and satellites. Since added signal strength is seldom needed in more than one place at a time to overcome rain attenuation, extra power could be provided in the satellite to penetrate heavy rainstorms, along with a mechanism for directing this power to the earth stations requiring it. A transmission technique called time division multiple access (TDMA) provides a means for increasing power during rainstorms, although no extra electrical power is used in the transmission. In TDMA, a single radio frequency is shared by multiple stations, with each station transmitting a burst of digital data during a specific time-slot.

New satellite systems will use the higher frequency transmission bands as the lower frequency bands become saturated. In using these higher frequency bands, TDMA transmission techniques will allow effective digital transmission between the satellite and the earth stations, even during atmospheric disturbances.

7.1.2 Satellite Communications Structure

A limited number of common carriers offer satellite communications services. Figure 7–2 shows the organizations that are directly responsible for the launch and control of satellite systems under their control, the common carriers using those satellite systems, and the areas of service for each satellite system.

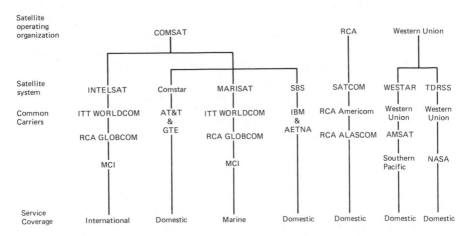

Fig. 7–2. Operational satellite systems.

The satellite-operating organizations are responsible for the development and continuing operation of their associated satellite systems. They provide common carriers with bulk satellite capacity. The common carriers sell satellite communication services directly to end-user organizations. The carriers are responsible for the development and implementation of the terrestrial stations. They are also responsible for the operation of terrestrial facilities connecting earth stations and for the earth station networks. They are generally not responsible for control of the satellite itself.

7.2 SELECTING A TRANSMISSION METHOD

Many users depend entirely on the telephone company to provide lines and equipment for all their data communication needs. Of course, the phone companies are the backbone of our communications network, but there are viable alternatives.

Bell's telephone lines for data transmission are available in many varieties. The type 3002 voice-grade private line is the most common for high-speed (2.4 kbit/s and greater) data traffic. Even the type 3002 channel is available in several flavors: It may have C-or D-types of conditioning or it may be unconditioned.

The unconditioned circuit usually contains repeaters at specific intervals, as well as loading coils and echo suppressors. In addition, local telephone companies, such as New York Telephone Company, offer what is called a straight copper circuit, which comprises pairs of copper wires terminated at each end by Bell 829 channel-interface devices.

This type of circuit is available only when the entire circuit is within the same telephone-company central office. In essence, the straight copper circuit is composed of two local loops connected at the central office. Since there are no repeaters and no loading coils in the signal path, the user can, in most cases, use the less-expensive limited-distance modems. Limited distance modems may also be used on standard type 3002 unconditioned channels, depending on the length of the circuit and the data rate. Modem manufacturer can supply the specific details.

Five types of C conditioning are available from Bell. C1 and C2 conditioned lines may be ordered for point-to-point, multipoint, and switched configurations. C3 conditioining, which is similar to C2 conditioning, applies only to private switched networks with a maximum of four trunks and two access lines in tandem.

C4 conditioning can be ordered in two-, three-, and four-point circuits only. C5 conditioning can be ordered only in point-to-point circuits and is similar to C2 conditioning. C5 conditioning is primarily intended for overseas channels. The 3002-type channel with C conditioning has specific limits on attenuation distortion and on envelope delay distortion that exceed those for a 3002 unconditioned channel (see Bell's technical reference Publication 41004 for exact specifications). Since Bell charges a premium per conditioning type per termination point, the telephone company is required to keep line impairments from exceeding the limits set in the tariff.

When transmitting digital data, it is necessary to transmit many state-changes per second. Attenuation distortion describes the extent to which a signal traveling through a channel is affected as the frequency of the signal changes. The higher the data rate, the more changes per second the channel must handle. The ability of a channel to handle these changes is called high-frequency response. Good high-frequency response is the same as low attenuation distortion. C conditioning improves the high-frequency response of the basic 3002 channel.

Envelope delay distortion occurs because the amount of signal delay is not constant at a particular frequency. High-frequency signals are delayed differently from low-frequency signals. Since data signals are transmitted by pulses, it is important that a channel preserve the pulse shape. In a channel with excessive delay distortion, various parts of the pulse arrive at different times and cause the reception of a distorted version of the original pulse.

Since data transmission involves successive pulses spaced closely over time, the envelope delay distortion causes pulses to spill over into each other. This phenomenon is called intersymbol interference. C conditioning limits the permissable amount of envelope delay distortion within specific frequency ranges.

D conditioning, available in two types, specifically limits noise and harmonic distortion. D1 conditioning is offered for point-to-point channels, and D2 for two-or three-point channels. Bell charges for D conditioning on a per-circuit-per-month basis rather than per termination point per month. Minimizing noise and harmonic distortion is critical to data transmission because these impairments interfere with the accurate reproduction of the transmitted pulse shapes. Once these impairments change the original signal shape, a modem receiver is unable to recreate the signal shape regardless of its sophistication. To combat this problem, modem manufacturers maximize the differences between various pulse shapes. A signal that uses four voltage levels to define 2 bits per pulse is more resistant to noise and harmonic distortion than one that uses eight levels to define 3 bits per pulse. High data rates such as 9.6 kbit/s on voice-grade circuits require the best protection against these impairments. Most 9.6 kbit/s modems operate satisfactorily with lines meeting only the basic 3002 specifications for noise and harmonic distortion. By minimizing these major impairments, modems can tolerate a greater number of other impairments, which makes D conditioning worthy.

Unlike the analog circuits, the telephone company digital data service (DDS) does not use modems, but receives the customer's digital data directly into a data service unit (DSU). The system handles data at rates from 2400 bps to 1.544 Mbps. DDS presently services about 100 cities nationally.

Digital services currently being offered are:

```
DDS = DATAPHONE DIGITAL SERVICE
TDS = TERRESTRIAL DIGITAL SERVICE (1.544 MBIT/S)
SDS = SATELLITE DIGITAL SERVICE (1.544 MBIT/S)
HSSDS = HIGH-SPEED SWITCHED DIGITAL SERVICE (1.544 MBIT/S SWITCHED)
CSDS = CIRCUIT-SWITCHED DIGITAL SERVICE (56 KBIT/S SWITCHED)
```

7.2.1 Private Alternatives

Private wire is usually a 20 gauge twisted pair cable that has been installed and is owned by the user. Since there are no restrictions on the signal strengths or voltages run through your own cable, modems can be used which are designed to transmit "brute force" signals requiring very simple transmitter and receiver circuitry in the modem. These simple units are classed as "line drivers", and typically sell in the under $200 range for surprisingly good performance from 0 to 19.2 KBS. The line cost is about 25 cents per foot for a four-wire twisted pair line; to this you have to add labor cost for installation time and termination connectors.

Fiber-optic local data links consisting of optical modems, multiplexers, and cable are now available at reasonable cost from Canoga Systems, RCA, Valtec, and others. As with customer owned copper cable, fiber optic cable may be installed only on your own premises, or with the permission of a utility if public thoroughfares are accessed. Some benefits of fiber optics systems are data rates in excess of 20 MHz and immunity to electrical interference as discussed in Chapter 1.

Fiber-optic cable is as low as 95 cents/foot and distances of 3000 feet are practical and with repeaters much longer distances can be realized. A major consideraton is the number of data channels that can be accommodated in one 1/8" diameter cable as compared to copper, particularly if cable ducts are crowded. Because fiber-optic cables are not susceptible to electromagnetic radiation, problems such as ground loops, crosstalk, and lightning interference are eliminated. No electrical signals are transmitted between equipment interconnected by the glass fibers, thereby eliminating the possibility of electrical surges or short circuits. Moreover, it is almost impossible to tap into detection, a security advantage over coaxial cable.

7.2.2 Measured Use Services

The DDD service we know best is the telephone company's direct distance dialing. This service charge is based on distance and connect time. If the data to be transmitted each week only amounts to an hour or two, DDD will be hard to beat.

Wide area telecommunications service (WATS) is available on either a measured time rate (10 hours per month included in the base rate; or full business day rate with 240 hours/month in the base rate. Overtime usage is accumulated in 6 minute increments. WATS would be a good choice if the requirement dictates contacting multiple locations on a fairly regular but not a full time basis.

TWX and Telex lines are used for data transmissions only and are limited to speeds of 150 and 50 bps respectively. DDS and WATS are offered by the telephone company; TWX and Telex are both offered by Western Union.

Specialized carriers, such as MCI Southern Pacific, and others offer measured use of their networks on a dial-in basis. For a low monthly fee ($10 to $35) you receive an access code and phone numbers in the various served cities. Usage is charged at a relatively low rate of 20 to 30 cents per minute regardless of distance. These services should be given serious consideration when the cities served fit your requirement.

Tymnet provides a "virtual connection" network accessed by DDD, WATS, or Tymnet provided access lines. Speeds are up to 4800 bps and rates are based on connect time, number of 1000 character block units, and any network interface equipment required.

Packet switching services like Telenet have transmission rates from 50 bps to 56 kbps. The service can be either dial accessed, or connected via dedicated line. Use charge is based on the number of kilo packets. Each kilo packet contains 128 characters.

The data communications planner should make a realistic estimate of actual data rate required, usage expected in connect time, the amount of data to be transferred, and the distance involved. These factors, properly evaluated, will allow selection of the most cost effective service for transmission.

7.3 FACSIMILE SERVICES

Facsimile services are generally available from one of the established common carrier companies and permit user access to their network for the purpose of transmitting fax copies. This can be a cost-effective alternative to designing one's own facsimile network in that the carrier company provides all that is needed in the way of switching and control hardware; the user needs only to have a facsimile terminal. Other advantages of the facsimile service come in the way of reduced line rates, or even distance-insensitive flat-rates covering the entire country. Some of the currently available facsimile services are included in the following paragraphs.

FAXPAK is operated by ITT domestic transmission systems. This service provides store-and-forward communications between compatible as well as many incompatible facsimile terminals. Available in the continental United States, the service is priced per minute on a "delivery priority" basis (15 minutes or 2 hours). This convenient structure offers lower sending costs for lower priority demands, and allows the user to make overnight transmissions at lower rates. Subscribers can access FAXPAK switching centers via analog or digital private lines or public dial-up lines.

Graphnet, a subsidiary of Graphic Scanning Corporation, is an authorized special common carrier that began its Fax-Gram facsimile service in January 1975. Available on a national basis now, the tariffed service is based on a value-added network that accepts, stores and forwards digitized graphics and messages through the company's computer/concentrators. The service permits input to be accepted from virtually any type of fax, CRT, Telex, or other keyboard terminal registered for use with the network. The rate structure offers three major classes of service: Class 1 and 2 for delivery of information to network facsimile devices during the normal business week and hours or after hours and on weekends and holidays; Class 3 for delivery to a Graphnet facsimile office and then by local messenger of pick-up by addressee.

Southern Pacific Communications, another common carrier company, introduced their Speedfax service at the beginning of 1978. The service provides a means of transmitting facsimile or other data between authorized facsimile and/or data terminals. The service is offered in 60 cities throughout the United States and is divided into six classes of operation: Class I employs SPC-provided 4-minute analog fax machines; Classes II, III, IV and V employ customer-provided analog and digital machines; and Class VI employs SPC-provided sub-minute digital fax terminals. All Speedfax communications are sent to the SPC data forwarding facilities which then performs any code conversions necessary and forwards the material to the respective address. Like FAXPAK, Speedfax provides various price ranges based upon a priority ranking for a message. Three priority choices are offered: Type A, under which transmission is attempted within 15 minutes; Type B, under which transmission is attempted within two hours; and Type C, next day delivery. The user is initially charged with a low subscription fee for each terminal. Transmission rates will vary by class of service and type of priority required. If the user chooses to employ an SPC-provided fax terminal, there will also be a monthly usage charge for the equipment.

MCI Telecommunications offers an international facsimile service for sending fax messages from the United States to overseas sites. MCI has five public facsimile centers—in New York City, Washington, D.C., San Francisco, Miami and New Orleans—from which the overseas transmissions can originate. The customer can send his document to one of these centers by facsimile, mail or messenger. One of the more attractive features of this service is that the customer does not have to own a facsimile unit; documents can actually be hand-delivered to the nearest MCI office.

RCA Global Communications offers a service called Q-Fax. The RCA service is also geared towards international fax communications. Documents can be delivered to an RCA Q-Fax operation center via facsimile or in person for over-the-counter service. Q-Fax centers are located in New York City, San Francisco, Washington, D.C., Honolulu, Guam and San Juan. Documents can be sent from these centers to more than 14 overseas locations.

GLOSSARY

Acoustic coupler. A form of low speed modem that sends and receives data using a conventional telephone handset and not requiring a permanent connection to the line.

Analog-to-digital (A–D) converter. A device which converts an analog input signal to a digital output signal carrying equivalent information.

Asynchronous system. A system employing start and stop elements for individual synchronization of each information character, or each word or block.

Attenuation. The difference (loss) between transmitted and received power due to transmission loss through equipment, lines or other communication devices.

Automatic calling unit (ACU). A device which, on receipt of addressing information from a business machine, automatically dials calls over the communication networks.

Automatic dialer. A device which will automatically dial telephone numbers on the network.

Automatic send-received (ASR). A type of teleprinter with auxiliary off-line storage that can enable the teleprinter to operate automatically.

Bandwidth. The difference between the highest and lowest frequencies of a channel, measured in cycles per second (Hertz).

Baseband.   The frequency band occupied by information-bearing signals before they are combined with a carrier in the modulation process.

Baud.   A unit of signaling speed. The number of signal elements per second where all such elements are of equal length and represent one or more information bits.

Bit.   A contraction of the term binary digit. A bit can be either 0 or 1 and is the smallest possible unit of information making up a character or word in digital code.

Blocking.   A condition where connections cannot be made due to "all lines busy".

Bridge.   Equipment used to match circuits to each other ensuring minimum transmission impairment. Bridging is normally required on multipoint data channels where several local loops or channels are interconnected.

Broadband.   Refers to transmission facilities whose bandwidth (range of frequencies they will handle) is greater than that available on voice grade facilities. Also called wideband.

Buffer.   A temporary storage device for data which cannot be used or retransmitted immediately.

Buffered network.   A data communications system which employs buffers associated with each terminal device to maximize the efficiency of terminals and facilities.

Buffered terminal.   A terminal containing a memory that stores input from the keyboard temporarily, until a transmit key is depressed.

Bus.   An electrical data path that serves as a common connection for a related group of devices.

Byte.   A sequence of normally 8 bits, operated on as a unit.

Carrier.   A high-frequency radio signal which is modulated to carry information long distances through space or via cable.

CCITT.   (Comite Consultatif International Telegraphique et Telephonique). The International Telegraph and Telephone Consultative Committee—an international organization concerned with devising and proposing recommendations for international telecommunications.

Central office (CO). The location of telephone switching equipment where customers' lines are terminated and interconnected. Also called switching center.

Central processing unit (CPU). A unit of a computer that includes circuits controlling the interpretation and execution of instructions.

Centrex. A type of private branch exchange service where incoming calls may be dialed direct to extensions without operator assistance. Out-going and intercom calls are dialed by extension users.

Circuit grade. The grades of circuits are broadband, voice, subvoice, and telegraph. Circuits are graded on the basic line speed expressed in characters per second, bits per second, or words per second.

Circuit switching. A method of communications where an electrical connection between calling and called stations is established on demand for exclusive use of the circuit until the connection is released.

Cluster. Refers to a group of terminal devices grouped in one specific location.

Codec. An assembly comprising an encoder and a decoder in the same equipment.

Common carrier. A company which dedicates its facilities to a public offering of universal communications services and which is subject to public utility regulations.

Conditioned circuit. A circuit which has conditioning equipment to obtain the desired characteristics for voice or data transmission.

Converter. A device which translates from one speed, code, or medium to another speed, code, or medium.

Crossbar. A type of common control switching system using the crossbar or coordinate switch.

Cross talk. An unwanted signal from one transmission circuit detected on another circuit.

Data access arrangement (DAA). A device provided by the telephone company, used to connect privately owned or customer-provided equipment (data sets) to the switched telephone network.

Data circuit. A means of two-way transmission between any two data terminating devices such as teleprinters or computers.

Data circuit-terminating equipment (DCE). The equipment installed at the user's premises which provides all the functions required to establish, maintain, and terminate a connection, the signal conversion, and coding between the data terminal equipment and the common carrier's line; e.g.; data set, modem.

Data switch. A location where an incoming data message is automatically or manually directed to one or more outgoing circuits, according to the intelligence contained in the message.

Data terminal equipment (DTE). Equipment at which a data communications path begins or ends.

Dedicated circuit. A circuit designated for exclusive use by a user.

Demodulator. A component of a data set or modem which is responsible for recovering data from received signals and converting to a form suitable for the receiving business machine.

Digital to analog (D/A) converter. A device which converts a digital input signal to an analog output signal carrying equivalent information.

Distribution frame. A structure with terminations for connecting the permanent wiring in such a manner that interconnection by "crossconnections" may be made readily.

Downlink. The communications link from the satellite to the receiving earth station.

EIA interface.   A standardized set of signal characteristics (time duration, voltage and current) specified by the Electronic Industries Association.

Error detecting code.   A code in which each data signal conforms to specific rules of construction so that departures from this can be automatically detected.

Facsimile.   Transmission of pictures, maps, diagrams, etc. The image is scanned at the transmitter, reconstructed at the receiving station, and duplicated on some form of paper of film.

Four-wire circuit.   A two-way circuit using two paths so arranged that the electrical signals are transmitted in one direction only by one path and in the other direction only by the other path.

Framing bit.   A binary digit which is used for data synchronization.

Frequency division multiplexing (FDM).   A multiplex system in which the available transmission frequency range is divided into narrower bands, each used for a separate channel. Channels are derived by allocating or "splitting up" a wider bandwidth into several narrower bandwidths.

Frequency modulation (FM).   The process of modifying the frequency of a carrier wave in step with the amplitude variations of the signal to be transmitted.

Frequency shift keying (FSK).   A commonly used method of frequency modulation in which a one and a zero (the two possible states) are each transmitted on separate frequencies.

Front end.   Device which interfaces the communications network to a computer's input/output channel.

Front-end processor.   A communications computer associated with a host computer. It may perform line control, message handling, code conversion, error control, and applications functions such as control and operation of special purpose terminals.

Full-duplex.   Refers to a communications system or equipment capable of transmission simultaneously in two directions.

Garble. An unintelligible received message.

Half-duplex circuit. A circuit which provides transmission alternately in either direction.

Host computer. A computer interconnected with a network that performs the computational jobs.

Host interface. An interface which connects a host computer with a communication processor.

Intelligent terminal. A terminal that contains a processing element. It can perform local data processing and storage activities.

Jitter. Short term line or circuit instability.

Keyboard send/receive (KSR). A teleprinter that enables an operator to send or receive data.

Leased line. A communication channel provided for the exclusive use of a customer, by a common carrier, at a fixed monthly rate.

Line printer. A printing device which prints an entire line of data at a time and then advances to the next line.

Local loop. The part of a communication circuit from the subscriber's equipment of the line-terminating equipment.

Mean time between failures (MTBF). A measure of reliability usually expressed as the average number of hours between one random failure and the next.

Message switch. A network node used to route data transactions to other nodes in the system.

Microwave links. Systems which use the relatively short microwave frequencies to broadcast from one point to another.

Modem. Contraction of the term modulator-demodulator. A device to convert one form of signal to another form for facility compatibility. For example, a modem is used to convert a digital signal from a computer into an analog signal so that it may be transmitted over the network.

Modulator. The equipment or apparatus which modifies some characteristic of a signal.

Monitor jack. A jack that provides access to communications circuits for the purpose of observing the signal conditions on the circuit without interrupting the service.

Mutiplexer. Electronic device for combining several signals into a composite stream for economic transmission. Techniques employed are frequency division (FDM) and time division (TDM).

Multipoint line. Refers to a single link connecting several terminals or nodes.

Narrowband. A channel whose bandwidth is less than that of a voice grade channel. Commonly used for communication at speeds of less than 300 bits per second.

Network. A series of points interconnected by communications channels, often on a switched basis. Networks are either common to all users or privately leased by a customer for his own use.

Nodes. Those points on a communications network at which a message can be entered into the network or received from it.

Packet assembler/disassembler (PAD). Equipment providing packet assembly and packet disassembly facilites.

Packet switching. A method of routing standard size groups of information called packets.

Patch panel. An arrangement for connecting channels and terminals in which plug cords are inserted into jacks corresponding to the channels or terminals to be interconnected.

Port. Entrance or access point to a computer, multiplexer, device, or network where signals may be supplied, extracted, or observed.

Pulse code modulation (PCM). The form of modulation in which the modulating signal is sampled, and coded, so that each element contains a different number of pulses and spaces.

Ring network.   A computer network in which each computer is interconnected to adjacent computers in a circular fashion.

Simplex.   A communication system capable of sending information in one direction only.

Specialized common carrier.   A common carrier offering limited services to customers in a particular geographic area.

Star network.   A communication system consisting of one central node with point-to-point links to several other nodes.

Status channel.   A channel to transmit status information.

Store and forward switching.   A method of transmission in which messages received at intermediate points are stored and then retransmitted to the intended destination.

Synchronous data network.   A data network in which the timing of all components of the network is controlled by a single timing source.

T-Carrier.   A series of highspeed transmission systems.

Teletypewriter.   A typewriter device capable of sending and receiving alphanumeric information over communications channels. Also known as a teleprinter.

Telex.   An automatic dial-up teletypewriter switching service provided on a worldwide basis by various common carriers.

Time division multiplexing (TDM).   A technique for combining several channels into one facility or transmission path in which each channel is allotted a specific position in the signal stream based upon time.

Trunk.   A large-capacity, long-distance channel used by a common carrier to transfer information between its customers.

Two-wire circuit.   A circuit formed by two metallic conductors insulated from each other. The term is also used, in contrast to a four-wire circuit, to indicate a circuit using one line or channel for communications in both directions.

Uplink.   The communications link from the transmitting earth station to the satellite.

Voice grade.   Refers to channels designed specifically for voice communication.

Wide area telephone service (WATS).   A flat rate long distance telephone service provided on an incoming or outgoing basis.

Wideband.   A system with a multi-channel bandwidth of 20 kHz or more.

The author credits the following sources for listings used in this glossary:

American National Standard Vocabulary for Information Processing, American National Standards Institute, Inc., June 1980.

Federal Telecom Standard 1037 Vocabulary for Telecommunications, May 1978.

INDEX

Access methods, 87
acknowledgments, 39, 40, 43
acoustic coupler, 19
advalanche photodiode, 12
aggregate, 25
alarming, 73
analog to digital conversion, 6
analog transmission, 6
ASCII code, 4, 44, 45
asynchronous, 8
auto answer device, 22

Backend network, 85
Bandsplitter, 31
bandwidth, 6, 24
baseband, 6
baud, 6, 9, 19
baudot, 44, 45
BCD, 43
bisync, 38, 39
bit stuffing, 41
breakout box, 77
buffers, 54
bus topology, 84

Cable choices, 89
CCITT, 20
channels, 4, 5, 21
circuit switching, 93, 94
coax, 91
codes, 43
Comsat, 110
concentrator, 25
conditioning, 111, 112

converters, 44
CRC error detection, 40, 47
CRT terminals, 51
CSMA/CD, 87
cursor, 53

Data monitor, 79
data service unit, 113
DCE, 33, 34
DDD, service, 114
deterministic network, 86
diagnostics, central, 71
dial backup, 22
digital transmission, 6
distortion, 112, 113
distributed network, 4
DTE, 33, 34

Earth station, 107
EBCDIC, 44
encryption, 80
error rate testers, 78
errors, 45
    control of, 39
    detection of, 40, 47
    effects of, 48
Ethernet, 87

Facsimile, 63, 64
facsimile services, 115
fallback, 69
fiber optic cable, 11, 90, 114
fiber optics, 9
frame, 42

frequency division, 24
frequency shift keying, 17, 24
front cnd network, 84
front end processor, 49
full duplex, 5, 56

Graded index fiber, 11
graphic terminals, 62, 63

Half duplex, 5
handshaking, 35
HDLC, 33, 41
horizontal distribution, 92

Idle character, 28
image resolution, 66
infrared emitting diode, 12
injection laser diode, 12

Jackfield, 69, 70

Keyboard, 53

Laser printer, 60
leased line, 93
line analyzer, 76
line costs, 14
line driver, 114
link control procedure, 33
local area network, 86
logical channel, 101
logical matrix, 105
loop check, 48
loop topology, 83

Master/Slave topology, 83
message switching, 93, 94
Modem, 7, 16, 17
    applications, 19
    features, 20
    voiceband, 19
    wideband, 19
Modulation, 17
    FAX, 64
    FSK, 17, 24
    multilevel, 6, 9
    PSK, 17
    QAM, 18

monitoring, 69
multidrop, 15, 55
multiplexers, 16, 22, 24
    bit oriented, 27
    character oriented, 27
    FDM, 24
    STAT, 25, 29
    TDM, 28, 29
    type comparisons, 26
multivendor network, 76

NAK, 43
network, 3
    backend, 85
    control center, 73
    design, 103
    frontend, 84
    security, 79, 80
    testing, 71
node, 100

Optical fiber, 90, 91
optical mark reader, 66

Packet assembler, 101
packet operation, 102
packet switching, 98
packet switching services, 115
PAD, 101
parallel transmission, 4
parity, 45
    double, 46
    spiral, 47
    vertical, 46
patch panel, 69
phase shift keying, 17
PIN photodiode, 12
polling, 43
printer, 56
    band, 59
    daisywheel, 57
    impact, 57
    line, 59
    matrix, 58
protected field, 54
protocol, 32
    comparison, 42
    hierarchy, 32
public data network, 96

Quadrature amplitude modulation, 18

Raster graphics, 62
read coding, 66
redundancy, 70, 106
repeaters, 6
ring topology, 83
RS449 interface, 20
RS232C interface, 20, 34

Satellite, 107
  channel, 107
  structure, 110
  systems, 110
scanning, 29
scrolling, 54
SDLC, 41
serial transmission, 4
simplex, 5
single mode fiber, 11
SNA, 42
star configuration, 82
statistical multiplexer, 25, 29
statistical network, 86
step index fiber, 11
storage tube, 62
stroke writing, 62
switching, 69

synchronous, 8
SYNC charactrer, 28

TDMA, 110
telex, 115
terminal considerations, 54
test equipment, specialized, 75
testers, portable, 75
time division multiplexing, 27, 28, 29
token passing topology, 83
topology, 82
traffic statistics, 104
transponder, 107
trouble tickets, 74
trunk testing, 74
twisted pair, 91
TWX, 115

Uplink, 109

Vertical distribution, 92
VF patching, 69
virtual call, 99
virtual circuit, 101

WATS, 114
workstation, 67

X.25 standard, 99, 100